SpringerBriefs in Applied Sciences and Technology

PoliMI SpringerBriefs

For further volumes:
http://www.springer.com/series/11159
http://www.polimi.it

Funda Atun

Improving Societal Resilience to Disasters

A Case Study of London's Transportation System

**POLITECNICO
DI MILANO**

 Springer

Funda Atun
Architecture and Urban Studies
Politecnico di Milano
Milan
Italy

ISSN 2282-2577   ISSN 2282-2585 (electronic)
ISBN 978-3-319-04653-2   ISBN 978-3-319-04654-9 (eBook)
DOI 10.1007/978-3-319-04654-9
Springer Cham Heidelberg New York Dordrecht London

Library of Congress Control Number: 2014937311

Printed on acid-free paper

Springer is part of Springer Science+Business Media (www.springer.com)

*I dedicate this book to
my beloved sister Zeren*

Foreword

In 2010 the UNISDR launched the "Making Cities Resilient" campaign, having as a goal to enhance the awareness of city mayors and managers about the impellent need to improve the capacity to prevent and respond much more effectively to the consequences of natural hazards. We are living in an urbanized world, for the first time in history the majority of the human population is living in cities and metropolitan areas. Many of the latter have been developed along coastlines, exposed to even higher threats of extreme storms, tsunamis, and hurricanes. These are the reasons that pushed the United Nations office to move from a general call for disaster risk reduction to more focused programs, the "Making Cities Resilient" campaign being certainly one of the most relevant in the recent years.

However, while raising awareness is crucial, it is not sufficient to achieve concrete results. There is the need to transform concepts and precepts into operational tools that can be applied in the field to lower existing vulnerabilities and enhance communities' coping capacity. While the literature offers many products on the theoretical aspects of resilience and vulnerability, few provide concrete experiences, methods, and indicators to measure resilience and vulnerability. The book by Funda Atun provides a really good example of how resilience can be tackled in practice in a large metropolitan area such as London. She has been able to examine the London situation through the eyes of an external observer who had the possibility to visit and stay for a while in the largest capital city of the European Union. Funda shows in the book how the concept of resilience can be operationalized at different scales, ranging from regional/metropolitan to local, and introduces the need for an innovative perspective according to which prevention goals must be sought at all scales and levels, involving the citizens at the local level.

In order to find the best indicators to use, at those different scales, Funda has drawn on the experience gained in the Ensure (Enhancing resilience of communities and territories facing natural and na-tech hazards) project funded by the European Commission within the 7 FP and thanks to the collaboration with the Flood Hazard Research Centre at Middlesex University that was partner to the project and hosted Funda for some months in London.

Operationalizing the resilience and vulnerability concept would be already a very relevant achievement per se; however, Funda went a further step in her research by applying an innovative methodology to the transportation sector that has been perhaps a little neglected so far. Despite the obvious fact that transportation networks play a crucial role in the everyday functioning of cities and also, even more, during emergencies, few researchers have been focusing on it. Funda is exploring and commenting on the state of the art of research in this particular domain in the introductory chapter. It has not been an easy task to fully integrate three main topics that are often dealt with independently in disaster studies: cities resilience and vulnerability, the functional and systemic vulnerabilities of lifelines and in particular of the transportation sector, and the organizational and social failures in fully appreciating the existing threats. What is new therefore is the blending of different perspectives, different bibliographic sources with documentation, and surveys gained in the field, so as to shed light on the potential impact of a flood on a critical infrastructure of one of the most important metropolitan areas in the world. The result is not just an interesting reading, but also a different perspective on risks that shows how the urban history of a city, its resilience, or lack of, have been shaped by a series of past incidents, and how today's level of risk results from a combination of "big" strategies and the many small decisions, the behavior of individuals with or without direct responsibilities in disaster management.

Scira Menoni

Preface

This book is a part of the Ph.D. thesis published in March 2013 at Politecnico di Milano. The thesis is about enhancing resilience of the transportation system against natural disasters and includes two case studies: London and Istanbul. The book provides you with the results of the London case study. I have tried as a foreigner to understand the current disaster risk management system in London. In addition to reading the literature and reports open to public, to observe the real situation and to produce primary data I decided to include interviews as one of the tools in the survey. Then, I found myself sending emails around and talking to people directly on the street. Actually, this is not a standard procedure in London; however, the result was surprising and I found institutions and individuals becoming very interested in the content of my work. The three-levelled approach of the survey was constructed during the study so as to see the links between organizational, tactical, and public levels in a society and effects of decisions coming from this interaction on the spatial pattern. The construction of the survey and accordingly the results of it are based on my perspective. It is very likely that as a foreigner I could not grasp the current situation entirely and I might miss some issues. For this reason, in addition to the insights acquired from the survey, the original questionnaires and the pure results are given at the end of the book. The results were attained in 7 months and provide an external perspective to the current disaster risk management system. It is my hope that this book will contribute positively to the current system in London.

Funda Atun

Acknowledgments

First, I am grateful to my supervisor, Prof. Scira Menoni, for her guidance and the clarity she has brought to my ideas. I cannot thank her enough for the opportunities she has provided me, and for truly looking out for my best interests.

Second, I am indebted to Prof. Dennis Parker and Prof. Sue Tapsell for hosting me for more than 6 months as a visiting researcher in the Flood Hazard Research Centre at Middlesex University in London. Additionally, I thank Prof. Dennis Parker for his great personal kindness, guidance, and many constructive comments throughout the development of the methodology and survey in London. I am also grateful to all members of the FHRC at Middlesex University for their cooperation and guidance.

I am pleased to acknowledge several others who contributed directly or indirectly to this research. Space does not permit to list the name of all the governmental and private institutions in London. More than 80 interviews had been conducted within this study, I have to thank all individuals and institutions that participated. This work would not be finalized without their precious efforts.

Chapter 1
Introduction

Abstract The book was set out to explore the ways to increase societal resilience in a complex city system against disasters by focusing on the transportation system of London. This chapter sets the background shortly and introduces the problems. Moreover, it outlines aims and objectives of the research. Finally, it provides a road map for the rest of the book.

Keywords London · Disaster resilience · Transportation system

> *London has always been a vast ocean in which survival is not certain.*
>
> Peter Ackroyd (2000)

1.1 Background

Since London was founded in A.D. 50, the city has experienced many disasters some of which destroyed almost the entire city and some others provoked extended damage (Withington 2010, p. 3). Both natural and man-made disasters, such as fires, floods, epidemics, wild weather, fog, wars, invasions, terrorism, train crashes, explosions and financial disasters occurred in the history of London. Following each disaster, London recovered, adapted to the changing environment and evolved in the light of the lessons learnt.

With its 8.2 million population (Office of National Statistics 2011) London is not the world's largest city any more, as it used to be in the 19th century. London is ranked the third largest city in population size in the European continent after Istanbul with 13.565.798 (TUIK 2011) and Moscow with 11.503.501 people (Russian Federal State Statistics Service 2010), and the biggest city within the borders of the European Union. However, with its 300,000 employees inside the borders of the Greater London Authority, and 800,000 employees in the region, London is one of the global centres for international banking and financial services by leading in foreign exchange turnover, trading of international bonds, trading overseas equities and fixing the gold prices (The Global Financial Centre Index March 2009). London precedes all other cities in all areas of competitiveness, closely followed by New York and Tokyo in the world and by Paris, Frankfurt and Zurich in Europe (The Global Financial Centre Index March 2009, p. 7). Therefore, consequences of an unexpected disaster would affect not only the city but also several other regions that are strongly connected to London.

1.2 Problem Definition

The problems perceived regarding disaster risk reduction procedures in complex cities, such as London, can be organized in three groups.

- Disregarding interdependency between system's components and among systems.
- Ignoring indirect and multiple hazards, which mainly depend on complexity of an environment.
- Dealing with the social structure of a city as if it is separated from the physical structure, whereas the social structure is strongly embedded within the spatial pattern.

The book refers mainly to the third problem by giving "interdependency of components" and "indirect effects of a disaster" preferential consideration. The social structure of a city is strongly embedded within the spatial pattern of streets, buildings and other infrastructures. Hence, the structural sub-systems cannot be thought separately from the social sub-system. Therefore, the challenge is to provide infrastructural solutions that consider social usage patterns and probability of human failures. For instance, after occurrence of a hazard if the main gas valves are not shut down immediately, a large fire may be ignited and lead to large number of fatalities. Moreover, not having an evacuation plan which responds to people's necessities can end up with many failures. Last but not least, delays in hazard forecasting could lead to un-sufficient warning lead-time.

1.3 Aims and Objectives

The general aim of the book is to reveal ways to enhance societal resilience in London against flooding by focusing on transportation. The concept of resilience was introduced at the beginning of the 1970s to indicate natural systems' capability for absorbing perturbations, preserving their structure and keeping the system functioning. The characteristics of a person or group in terms of their capacity to anticipate, cope with, resist and recover from the impacts of natural hazards (Blakie et al. 2004). In ARMONIA project (2006; cited in Walker et al. 2011, p. 17) 'resilience' is defined as "the capacity of a system, community or society potentially exposed to hazards to adapt, by resisting or changing in order to restore or maintain an expectable level of functioning and structure."

Although the ultimate **aim** is to gain understanding on how to increase societal resilience, the survey starts with apparently simpler questions: "what are the awareness and preparedness of risk at the organizational, tactical and public levels?" and "what are the effects of the outcomes of the decisions coming from organizational, tactical and public levels on the transportation system?".

The objectives are understanding horizontal and vertical interactions among parts, and as a result, understanding contribution of interactions to behaviour of the larger system. As it has been stated before, the book focuses on structural and operational (organizational and tactical) connections among elements as well as on elements themselves, to gain a better understanding of how an entire city system functions and how agents react during an emergency.

By following this approach, the structure of the book is organized to

- understand structural and operational connections among parts of the systems.
- understand how awareness and behaviour of different social groups shape the response of transportation system during a disaster.

1.4 The Transportation System as the Core Subject

When it is asked what the basic needs of people are, one could say that food, clothing and shelter. In the classic article "Maker and Breaker of Cities", Clark (1958) says that these are just the end-products and it is impossible to produce them without presence of a transportation system. Clark explains vital importance of a transportation system, and relationship between a well-developed transportation system and increased size of a city. The transportation system is vitally important to keep the city system functioning, as it is strongly interdependent to other sub-systems.

The reasons of choosing the transportation system as the core subject are listed as in the following:

- The transportation infrastructure often requires longer repair time with respect to other lifelines.
- Failures in the transportation system could stress pre-existing conditions of vulnerability.
- The transportation system is interdependent with many other systems, such as water, gas, sewerage, electricity, telecommunication, fuel supply, firefighting, structurally and/or functionally.
- Absence of the transportation system could have economic impacts on the long run. Enterprises depending on the transportation system can be closed or moved to another location, because the transportation system needs longer restoration times than other systems.

- During an emergency the transportation system is one of the most important systems to respond and keep a city system functioning. Failures in keeping the transportation system functioning may lead to increased death toll.

1.5 The Structure of the Book

The book begins theoretically with a literature review mainly focusing on the spatial evaluation of London by considering housing and transportation systems and natural disasters. Then, London is analysed to understand the nature of transportation system and how it may react in case of flooding. Both qualitative and quantitative analyses are adapted to London.

Understanding the specific local conditions and the context related features are crucial for any intervention in a complex city system. This book is organized around three main parts: the historical part, the current situation and a final re-conceptualization of the problem, based on the empirical work carried out in the field. Those three main parts are organized in six chapters.

In order to obtain an introductory outlook, Chap. 2 starts with a brief history of London by focusing on housing and transportation systems to analyse root causes of current vulnerabilities to disasters. Then the chapter proceeds by providing retrospective view of risk in London including both natural and man-made disasters. The second chapter refers to various events to understand how disaster risk reduction system has evolved and reached to today's level.

Chapter 3 includes the current situation in London, and aims to understand today's London including the role of transportation, present flood risk and emergency planning systems.

Chapter 4 reveals current systemic and social vulnerabilities by focusing on the downstream of the River Thames.

Chapter 5 describes the way that the survey was designed and how interviews were conducted with the stakeholders from the organizational, tactical and public levels. Moreover, the chapter includes the survey in three steps including people from the organizational, tactical, and public levels. Aims of the survey are to understand preparedness of organizational, tactical and public levels, perception and awareness of risk condition at the organizational, tactical and public levels, and awareness of flood warning, access to information programmes and individual preparation at the public level. The list of the questions and detailed results of the survey are given in Appendices A, B and C.

Chapter 6 covers discussion and conclusion of the survey. The discussion part comprises relationships among results and generalizations, the relationship of present results to research questions. Besides, new observations, new interpretations, and new insights that have resulting from the research are given in the conclusion sub-section.

References

Ackroyd P (2000) London: the biography. Chatto and Windus, London

Blaikie P, Cannon T, Davis I, Wisner B (2004) At risk. Natural hazards, people's vulnerability and disasters, 2nd edn. Routledge, London

Clark C (1958) Maker and breaker of cities. Town Plann Rev 28(4):237–250

Office of National Statistics (2011) Census on 27 March 2011. http://www.ons.gov.uk/ons/rel/mro/news-release/census-result-shows-increase-in-population-of-london-as-it-tops-8-million/censuslondonnr0712.html. Accessed Jan 2014

Russian Federal State Statistics Service (2010) Population census in 2010 (in Russian). http://www.gks.ru/free_doc/new_site/perepis2010/croc/perepis_itogi1612.htm. Accessed Jan 2014

The Global Financial Centre Index (2009) March 2009. http://www.romandie.com/news/pdf/PDF_Classement_GFCI_des_places_financis__ZH_5_GE_6_ROM_200320120403.pdf. Accessed Dec 2012

TUIK (2011) Turkish Statistical Institute, census on 2 October 2011. http://www.tuik.gov.tr/PreHaberBultenleri.do?id=15843. Accessed Jan 2014

Walker et al (2011) Introduction to sustainable risk mitigation for a more resilient Europe. In: Menoni S, Margottini C (eds) Inside risk: a strategy for sustainable risk mitigation. Springer, Berlin

Withington J (2010) London's disasters. History Press, Stroud

Chapter 2
Retrospective View of London

Abstract A brief history of London is given by focusing on changes on the spatial pattern regarding housing and transportation systems to analyse root causes of the current vulnerabilities to disasters. Then, the chapter proceeds by providing retrospective view of risk in London including both natural and man-made disasters to see the progress of disaster risk reduction system after each event in London.

Keywords Urbanization history of london · Disaster history of london · Transportation system

2.1 London's History as an Urban Centre

The population of England and Wales doubled between 1841 and 1901 rising from 15,914,88 to 32,527,843 people (The UK Registrar General Census 1961, p. 75). The effect of industrialization in a society that was living on agriculture just half a century before was drastic both in rural areas, which lost majority of their population by the second half of the nineteenth century, and in cities, as a consequence of rapid unstructured urbanization and poor living conditions of low income labourers (Banks 1968, p. 277).

Indeed, London was growing in size and population, and consequently, at the beginning of the nineteenth century, London had become an industrial and administrative centre. Besides, because of the increasing international sea trade, it became the busiest port in the world. While density of the city was increasing in the inner London, the impounded dock systems with warehouses were built on the both sides of the River Thames to handle the cargo system (Gilbert and Horner 1985, p. 7). The development of London was unstructured and did not consider the effects of development on the entire city pattern, as there was no authority responsible for the growth of London as a whole before the establishment of Metropolitan Board in 1855 (Owen 1982, pp. 31–46).

Each historical perspective tells much about the nature of interventions in planning discipline, though, it makes sense to start from the Victorian city and investigate changing physical, economic and social patterns, and how those changes affect physical, social and economic vulnerabilities to disasters in today's London. London's history as an urban centre has been analysed in three sections: housing, urban regeneration in Docklands and transportation system.

2.1.1 Housing

2.1.1.1 Victorian City

In the Victorian London, overcrowded and disorganized clusters formed slums inhabited by low-income labourers, due to not having sufficient and affordable housing supply in the central London, and a well-developed and affordable transportation system that could let people to move outside the city (Hall 1988). The problem's root cause was economic, people had gathered in one area, as they could not afford to move out (Hall 1988).

Several state interventions and acts were issued to solve the problem, such as the Torrens Act (1868 The Artisans and Labourer Dwelling Act) and the Cross Act (1875 The Artisans and Labourer Dwelling Improvement Act) (Hall 1988). The one in 1868 led local authorities to build new dwellings for working classes and the latter gave the permission to local authorities to clear large areas for re-housing inhabitants (Hall 1988). However, the problems were not solved, because of the severe budget problem of local authorities.

In 1900, London was ranked as the biggest city in the world with its 6.5 million population; followed by New York and Paris (Chandler 1987). In the early twentieth century, the main concern of authorities in London was still housing. At last, the Royal Commission attempted to solve the housing problem by issuing a series of acts in 1888, 1890 and 1900, which provided a framework to redevelop large areas with compulsory purchase, and enable local authorities to buy land outside their own borders as well (Hall 1988). These acts, with the help of the new transportation technologies, led to first suburban settlements.

2.1.1.2 Suburbanization

Improved transportation system, a combination of private and large-scale urban development, and cheap long-term mortgages allowed development of suburbs and gave a new shape to London.

Unwin (Hall 1988) stated that in the twentieth century's London local authorities must build on cheap undeveloped lands in the periphery of the city by connecting them with tramways. Besides, Unwin defined his ideas about how physical characteristics of a healthy town should be in his *pamphlet* "Nothing gained by overcrowding": minimum distance between two houses should be 70 feet at least to guarantee winter sunshine, short terraces, back-land as recreational space, cul-de-sac to create safer places for children (Hall 1988).

However, applying suburbanization as a solution to the unhealthy structure of the city had posed significant problems. Hereafter, the city resulted in more dispersed, de-concentrated, and fragmented areas filled with mass produced and low-quality houses (Hall 1988). Moreover, suburbs were not providing job opportunities to people as cities, as a result, suburbs were not successful at first. In the Victorian City, people were living in poor conditions; however, the city was offering economic and social opportunities (Hall 1988).

Despite theoretical and legislation efforts though, in 1920 there were still 184,000 people living in unhealthy areas and 549,000 people in unsatisfactory conditions (Hall 1988). There were two remedies only: to build up and to move out. For this reason, garden cities for 30,000–50,000 people encircled by a green belt were seen as the long-term solution (Hall 1988).

2.1.1.3 The Garden City

The green belt concept suggested first in 1927 by Frank Pick (Hall 1988). He commented on the unplanned growth of London by saying "there was much planning, but no plan" (Hall 1988). He continued by suggesting to restrain the growth of London with a green belt at least one mile around London, in order to control new industries at the city's edge (Hall 1988), to preserve the natural land and to limit the urban spread (Correll et al. 1978, p. 207).

Ebenezer Howard stood with the idea of garden cities for an ordinary human being's basic interests and desires in matter of physical environment, and that was his first effective voice in showing the way to meet those interests and desires with the help of social actions (Osborn 1950, p. 221). For the first time in the history of urban planning, social issues and physical issues were considered together (Hall 1988). According to Hall (1988), Howard dreamed of voluntary self-governing communities. Howard took his main ideas for the garden city from the James Silk Buckingham's plan for the model town (Hall 1988). However, he moved away from an utopian view, and imagined a city where industries, farms and urban institutions were integrated.

The key points of Howards' garden cities were "local management" and "self-government". The main problem to achieve these key points was to coordinate housing and industrial developments (Hall 1988). In 1938, Howard's garden city was able to be completed on a rather smaller scale than originally planned. At the end, the garden city became the victim of land-speculations (Hall 1988).

2.1.1.4 Greater London Plan 1944

After the Second World War, due to obsolete housing stock and declined quality of urban life, large numbers of Londoners migrated to suburbs. To solve the problems regarding housing, population sprawl and traffic congestion, Patrick Abercrombie's plan for London's development was prepared between 1942 and 1944. The intention of the plan was to surround London with a green belt and regroup the population both in the new and enlarged towns offering also workplaces (Parker 1999; Hall 1989, 2002). According to Hall (1989), the plan of Abercrombie was an interpretation of Howard's garden cities but at a bigger scale. By implementing the plan, the population moved beyond the green belt successfully and the city's physical sprawl stopped. However, due to an unexpected birth rate between 1955 and 1964, and increasing job opportunities in and around London the population of London increased more than the foreseen by Abercrombie and Barlow (Hall 1989).

2.1.2   Urban Regeneration and Docklands

The new government, established in 1979, tended to do minimalist planning, and replaced the main report prepared in 1970 with a three pages paper. The main concerns of the new government were "to promote economic recovery, to restrain public expenditure, to stimulate the private sector, to sweep away obstacles to commercial enterprise, to achieve more homeownership and housing in sale" (Hall 1989). To achieve these aims, deprived areas in the city have been chosen as the new redevelopment areas, such as Docklands.

During the aerial bombardment at the Second World War, London was damaged devastatingly, especially the working class houses in the East End and the Docks. Parker (1999, p. 193) indicates that "the Docks never fully recovered from this damage and their extinction was sealed by technological changes in shipping that began during the 1960s. Thereafter the use of bigger ships and containerization shifted port activities downstream away from the immediate vicinity of London".

These areas had been the object of an increasing interest during the 1960s. However, the depression in the 1970s changed the perceived problem in the planning discipline. By the late 1970s, large tracts of empty or semi-empty land were noticeable throughout Britain cities and ruins of disused industrial and warehouse buildings awaited redevelopment, such as the Docks (Hall 1988). Provision was urgently needed due to having large vacant industrial areas in the city centre.

The London Docklands Development Corporation had established in 1981 to revitalize the area by the State's Secretary for Environment. Creating an enterprise zone was seen as a solution. Therefore, in 1982 the Docklands became an enterprise zone where businesses were free from property taxes. Other initiatives, including simplified planning and capital allowances have increased the attractiveness of the area. The declined area converted into a mixed used zone that consisted of residential, commercial, and light industrial space (Hall 1988).

The period of Margaret Thatcher's Conservative political revolution (1979–1990) had witnessed abandonment of the strategic planning concept and dismissal of the Greater London Council in 1986, which had been the London's governing body since 1965. As a result, commercial enterprise potential was all at hand to shape London's future. Urban enterprise zones were created, and urban development corporations were set up to make necessary public investments that would bring private capital to the inner-city (Parker 1999, p. 198).

Between 1981 and 1990, invested public funds exceeded six times mostly for constructing office spaces; 41,421 new jobs had been provided in the Docks, together with sites for thousands of new homes (Hall 2002, p. 398). The Docklands' regeneration project reflects how private property intensely operated for achieving public ends (Hall 2002, p. 398). This domination had been fostered by the deregulation of British financial services in 1986 and had been coupled with an unexpected recovery of population growth in the central London (Champion 1987; cited in Parker 1999, p. 199).

With the help of the development program, Margaret Thatcher's conservative government had promoted the market-driven method regeneration in 1980 for the both sides of the River Thames. Consequently, middle class population took the place of formerly working class (Brownill 1990; Foster 1999; cited in Butler 2007, p. 760).

According to Slater, as quoted in Butler (2007, p. 760), this urban regeneration process took the form of developer-led gentrification. Afterwards, the Docklands model was extended to the rest of the Thames Gateway for supporting London's continued growth with the support of Greater London Council. According to Butler (2007, p. 760), this policy could help to stop the unsustainable suburban expansion by attracting people back into the city.

2.1.3 Transportation System

In the nineteenth century, London was the scene of many innovations on transportation systems due to increasing demand; London built the world's first public railway, the first underwater tunnel and the first underground railway. The world's first public railway was opened in 1803 near London.

By the 1830s, new railways were built and operated. London and Birmingham Railways were opened in 1837, and central London can be reached by the construction of the Euston central train station. London-Deptford railway was opened in 1836 and extended until Greenwich in 1838. In 1884 underground railway had been extended to inner suburbs of London (Hall 1989). With all these accomplishments, transport had become available also around the centre of London. By the end of the nineteenth century, different modes of transportation were connected to each other, and thus constructed a network of central and suburban railways and freight railways. By 1900, London had fifteen main railway lines (Fig. 2.1).

Although, by the mid-1850s, 27,000 people were commuting into London by train every day, 90 % of the workers were still going to work by walking or by omnibus (London Transportation Museum). Hence, companies started offering affordable tickets for the working class. In 1855, the first horse bus network had been developed by the London General Omnibus Company and was extended by the horse trams to reach the inner suburbs in 1870 (Hall 1989).

Fig. 2.1 London railways
by 1900 (Author's archive.
Modified *photo* from the
London Transportation
Museum, May 2011)

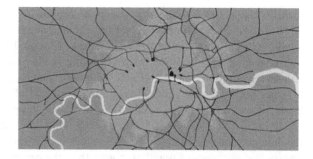

Fig. 2.2 Thames Tunnel
(Author's archive,
December 2011)

   An important obstacle to the development of the transportation system was
constituted by the River Thames, which was a barrier for the Docks and industries
on the both sides. More bridges and crossings were necessary. By 1880s, there
were only three bridges on the river: London Bridge (1200), Westminster (1750)
and Blackfriars (1769). By 1895 fifteen road and footbridges and 4 Railway
bridges were built over the River Thames. Moreover, Mark Brunel put forward the
idea of underwater crossing and the Thames Tunnel was constructed between
1825–1843 (Fig. 2.2). Today the tunnel is still in use by the London over-ground,
where the Rotherhithe and Wapping stations can be found (Fig. 2.3).
   After the Second World War, according to Patrick Abercrombie, London was
like a collection of villages without clear boundaries and with strong centres, and
they need to be connected by new arterial and sub-arterial highways hierarchically
(Forshaw and Abercrombie 1943, pp. 2–50). Besides, doing this could easily define
the edges of London and solve the traffic congestion problem (Hall 1989). In the
second half of the twentieth century, with the increasing suburbanization trend,
number of cars increased dramatically and traffic congestion became a huge
problem especially in the historical city centre. The Minister of Transport published

Fig. 2.3 Underground
station of Rotherhithe
(Author's archive,
December 2011)

a transportation report in Britain at the end of the 1963. Colin Buchanan stated
"planner should set fixed standards for the urban environment, whereupon more
traffic could be accommodated only through massive reconstruction; if the community were unwilling or unable to foot this bill than it must restrain the traffic"
(Hall 1988). Starting from the 60s Ministry of Transport was searching the best road
pricing system to restrain people from making journeys with their cars and
encourage public transportation. Smeed Report in 1964, the study of "Better Towns
with Less Traffic" in 1967, the research of London Planning Advisory Committee
(LPAC) in 1980 and the Road Charging Options for London (ROCOL) in 1998
provided various methodologies for road pricing to reduce the traffic congestion
(Dix 2002, pp. 2–3). Finally, the congestion charging scheme started in 17 February
2003. According to the results reported by TfL in 2008/09, the scheme succeeded to
decrease the traffic in the central London by 21 % (TfL 2009, p. 23).[1]

2.2 Retrospective View of Risk

London has reached today's advanced level at disaster risk management by
experiencing several types of disasters, both natural (floods, snow storms, ash
crises) and man-made (fire, wars and terrorist attacks) disasters during its history.
Herewith, some of the remarkable disasters, which have shaped current disaster
risk management system have been analysed by considering the lessons learned
and their effects on the structural/non-structural mitigation measures, forecasting,
monitoring and emergency response systems.[2]

This section focuses on the remarkable disasters in London's history, such as
1666 Great Fire, the air raids during the World War 2, 18 December 1987 Kings

---

[1] TfL, 2008: Congestion charging five years on. Available at: http://www.tfl.gov.uk/corporate/
media/newscentre/archive/8948.aspx.

[2] This section of the second chapter is published in Atun 2012.

Cross Fire, the terrorist attack to tube network on the 7th of July 2005, 1928 flooding and 1953 Storm Surge.

The Great Fire in 1666 and the air raids during the World War 2 provided the opportunity to shape the physical structure of London, as the biggest portion of the city was destroyed by those disasters (The first plan of London was issued after the Great Fire in 1966, and the second after the World War 2, by Abercrombie). The Kings Cross Fire and the July 2005 terrorist attacks revealed the strategic, operational and systemic problems encountered during an incident on the transportation system. Moreover, the 6 January 1928 flooding and the 1953 East Coast Surge floods are chosen as they had noteworthy effects on structural and non-structural defences to flooding in London.

2.2.1 Fires: 1666 Great Fire and 18 November 1987 King's Cross Fire

2.2.1.1 1666 Great Fire

While the Great Fire of London was a disaster destroyed almost the entire city, a much safer city was achieved by rebuilding according to the new rules defined by the "rebuilding of London act 1666". In addition, this event led to initiate professional fire fighting in London.

When the Great Fire occurred in 1666, London was the biggest city in the UK with its estimated 500,000 population (Withington 2010, p. 71). The Great Fire lasted 5 days and destroyed more than 436 acres of urban land. One in every three houses was destroyed by the fire and around 70,000 people (representing 14 % of total population) became homeless (Withington 2010, p. 71). After the incident, Sir Christopher Wren prepared the first plan for London, in which central streets provide connections between public squares and landmarks, while narrower streets divide residual areas according to a grid shape.[3] Moreover, fire regulations enhanced and "rebuilding of London Act 1666" was issued. This act required to widen roads and imposed regulation setting the type of structure that were allowed and the provision of a fire court. According to the Act, all the buildings had to be in brick or stone.

Permitted buildings were grouped in four categories:

- On smaller streets: cellar, two floors high with an attic on by-lanes.
- On larger streets: one more storey than the first category.
- On main roads: two more storeys than the first category.

---

[3] Royal Institute of British Architects (RIBA) http://www.architecture.com/LibraryDrawings AndPhotographs/OnlineWorkshops/UrbanAdventures/01Wren.aspx (accessed on 30.04.2012).

- Mansions with fewer restrictions than the other three but still restricted to four storeys plus cellar and attic.[4]

In accordance with the fire regulations, the city was divided into four districts. In each district, there were "800 buckets and 50 ladders, as well as shovels, pick axes and hand-held squirts"; people were also informed about how to use the fire-fighting equipment and how to quell a chimney fire (Withington 2010, p. 75). Moreover, fire insurance concept aroused, as many business people bankrupted due to the fire. First insurances were offered in 1680, by promising clients "the services of watermen as fire-fighters, or the rebuilding of their premises if these efforts failed to serve them", creating the first professional firefighting service in London (Withington 2010, pp. 75–76).

2.2.1.2   18 November 1987 King's Cross Fire

The fire at the King's Cross tube station was initially a minor incident, which could have had different results. However, it turned into a disaster, as a result of the chain of events, which occur regarding to employees' limited knowledge on distinguishing a fire and not well handled evacuation procedure.

The chain of events occurred during the incident indicates two key problems: untrained underground staff and deficiency of communication

- **Untrained underground staff** During the incident there were 23 staff at the station, however, only four of them received training in evacuation/fire drills (Withington 2010, pp. 104–106). Even though there were fire-fighting equipment and tools everywhere in the station, the staff at the station was not able to use them (Withington 2010, pp. 104–106).
- **Deficient communication** Communication was limited, as it would occur only through telephone or word of mouth. In this case, the supervisor of the station was in his room that was far from the fire, and the only way to communicate with him was his internal phone and he was informed about the fire 12 min after its first discovery (Withington 2010, pp. 104–106). Although fire brigades and British Police officers had radios, they were working only on the surface level (Withington 2010, pp. 104–106).

In this incident, more than 200 firemen were involved, 31 people died and more than 50 people were injured. Although, there had been about 400 fires in the London underground between 1956 and 1987 (Withington 2010, p. 98), the obligation of doing drills was introduced only after this tragic fire at the King's Cross tube station.[5]

---

[4] http://london.allinfo-about.com/features/rebuilding.html (Stephen Inwood, A History of London, accessed on 30.04.2012).

[5] Interview with the former CEO of TfL.

2.2.2 War: The World War 2 and Air Raids

In the second World War, in 1940, especially road networks, the Docks and railroads were bombed with the air raids. Withington (2010, p. 24) states that telegraph poles caught fire due to the heat after bombing. As fire engines were short in supply, taxis and private cars were used to carry mobile fire pumps. Even though their homes stand, people were out of water, gas, electricity, food and basic services. Moreover, sewage breached and contaminated potable water. Buses were used to evacuate people to rest centres in safer places. However, in the confusion, some of these buses could not find their way and could not arrive at safer places and rest centres. Regarding the tube network, people, who got stuck in London, used nearly 80 tube stations as shelters (Withington 2010, p. 27). Even though tube stations could seem as the best shelter, they were actually hit by bombs and hundreds of people died there.

In the first 6 weeks of the air raids, more than 6,000 people were killed and 10,000 injured. 16,000 houses destroyed, 60,000 seriously damaged and 300,000 people needed re-housing (Withington 2010, p. 28).

The London blitz provided the reason to plan and reconstruct the city again, as it happened after the Great Fire. Patrick Abercrombie prepared "the County of London Plan in 1943", which states the deficiencies of London, and "the Greater London Plan in 1944", which provides strategies and policies regarding deficiencies stated in the previous plan.

2.2.3 Terrorism: Terrorist Attack to Tube Network on the 7th of July 2005

In 2005, the terrorist attack to the tube network indicated what worked effectively and revealed the defects to be improved for better performance of the system.[6] When the bomb exploded at 8:51 a.m., it took time for the officers to understand what had happened. The first thing that was seen was a massive loss of electrical power on the northern side of the circle line. Two years before this event, on 23 August 2003, power loss was experienced in the London underground, and as a result, a dramatic incident occurred, when many people remained trapped in tunnels and trains. Therefore, as the first reports were saying that there was no electrical power, the initial assumption of the authorities was that a power failure had occurred again. Further reports showed that the situation was more serious and different than the initial assumption. By 9:15 a.m., it was decided to evacuate the tube network and the code amber[7] was ordered to ensure safety of employees, public and further trains.

---

[6] This part consists of the results of the interview with the former CEO of TfL.

[7] Code amber is the code to evacuate network and control under circumstances.

(according to values in 1953) (Jonkman and Kelman 2005, p. 2). Moreover, in the Netherlands 200,000 hectares area were inundated, in the UK 40,000 hectares and in Belgium 10,000 hectares (Jonkman and Kelman 2005, p. 2).

The causalities in the UK occurred due to forecasting failures. The event was unexpected and there was no warning. Fatalities were higher in the sea towns in Canvey Island, Jaywick and Lynn, where people started living in those kind of temporary and low quality buildings shortly after the war (Jonkman and Kelman 2005, p. 6). Main causalities were among elderly people. At Canvey Island 42 out of 58, at Jaywick 28 out of 34, and at Lynn all 14 fatalities were older than 60 years old (Jonkman and Kelman 2005, pp. 5–6).

Met Office indicated that the surge expanded from Tilbury to the Docklands and caused damage to the Docks, oil refineries, factories, cement works, gas works and electricity generating stations. Additionally, 100 m sea walls were destroyed, and, more than 1000 houses flooded.[8]

The shortcomings of the forecasting and warning systems in the UK became evident with this event. People were unaware of their own vulnerability to storm surges, and being without electricity and communication systems increased the number of causalities. Collapsed Sea defences raised concern about the maintenance of structural defences and their reliability (Jonkman and Kelman 2005, pp. 8–9).

2.3 Learning from Experiences and Changing Safety Measures

After the 1666 Great Fire, the first city plan was prepared to rebuild the city by considering mitigation of fire risk for the first time in London's history. Moreover, after the WW2, "Greater London Plan" had been prepared by the team of Sir Patrick Abercrombie to rebuild the city and to improve the existing deficiencies lasted since the beginning of the nineteenth century. The "Greater London Plan" has given London its current shape. In addition, the Docks, which were severely damaged during the blitz, was subject to regeneration and economic revival in the 80s, and today the area consists of residential and financial activities.

Furthermore, the fire at the King's Cross underground station revealed the deficiencies of communication and un-trained underground staff in case of an emergency. After this event, getting involved in drills has become obligatory for the underground staff. However, communication system was not improved until the terrorist attack in 2005. Terrorist attack on the 7th of July 2005 revealed the strategic, operational and systemic failures encountered during a disaster on the transportation system. Foremost problem in the latter was communication between

---

[8] Met Office: Great weather events: the UK east coast floods of 1953, http://www.metlink.org/pdf/teachers/1953_east_coast_floods.pdf.

Table 2.1 Retrospective view of risk (Atun 2012)

Event	Description	Deficiencies shortcomings	Effects on structural pattern	Effects on non-structural pattern
1666 Great Fire	Lasted 5 days and 436 acres burned. 70,000 people became homeless (14 % of London's total population)	Timber houses, narrow streets, not having sufficient equipment to extinguish fire	A grid plan prepared for rebuilding the city of London. "Rebuilding of London Act" issued, and all the buildings rebuild by brick or stone. Moreover, the act grouped the buildings in four categories and height of the buildings had arranged according to the width of the street	Fire regulations had improved, and the event led to initiate professional firefighting. The rebuilding act divide the city into four districts and sufficient firefighting equipment had provided for each district. People had informed about how to use the equipment to extinguish fire
Air Raids during the World War 2	More than 6,000 people were killed, 10,000 people injured, 16,000 houses damaged, 60,000 houses seriously damaged, 300,000 people needed re-housing	Fire engines were short in supply. Evacuation of people was not successful. Shelters were in shortage	The plan of Abercrombie was prepared to reconstruct the city again. The Docks were never fully repaired after the blitz and they were the regeneration areas during 80s	
1987 King's Cross Fire	31 people died, 50 people were injured, 200 fire brigades had been involved	Not having trained underground staff and communication problem	Timber escalators were removed	Doing drills has become obligatory for the employees of the TfL

(continued)

Table 2.1 (continued)

Event	Description	Deficiencies shortcomings	Effects on structural pattern	Effects on non-structural pattern
2005 Terrorist Attack to the Tube Network	Multiple attacks on the tube network	It took time to understand the real issue, as communication was a problem Being disciplined about following the protocol Not having medical supplies in the underground Employees knew what to do, because they were doing drills twice a year, however they never practiced a drill for multiple accidents	Stores of medical supplies were put across the system and very large supplies at the strategic locations in the zone one Inner operable radio system had built, know it is possible to talk to anyone no matter in the tunnel or not	To keep the trust with their employees, underground has changed their protocol and decided to communicate their employees and tell everything they know and do not know Having multiple attacks changed the scenario planning for drills
January 1928, Flood	The water was 1.8 m higher than predicted in the central London and 0.3 m more than formerly recorded level	Not having a proper forecasting and warning system Not having sufficient structural tools to prevent flooding	The height of the walls along the Thames had risen	Improvements on the forecasting and warning system: a research program for forecasting was started, and a warning system for London was established
East Coast Surge Flood 1953	More than 420 people died, 32,000 people were affected, and economic damage was 50,000 million pounds (1953 values)	The forecasting and warning systems were the shortcomings of this event The maintenance degree and reliability of structural defences were other problems	Thames Barrier had been built Structural defences had been improved	Flood forecasting, monitoring and warning systems have been improved Emergency management system has come to today's level

the staff in the tunnels and the people who were outside and trying to understand the real issue. As employees of TfL were doing drills since the King's Cross fire, they knew what to do in case of an emergency, and this was one of the biggest advantages. After the incident, a radio system was installed and communication has not been a problem anymore.

Regarding to flooding, after the flood event in 1928, the focus was on deficiencies of structural mitigation tools. Besides, attention was given also to build forecasting and warning systems by conducting a research program for forecasting and by establishing a warning system for London. However, in 1953, a surge flood hit the southern part of the country including London, and the event showed that the forecasting and warning systems, which were established more than two decades before, were not successful (Table 2.1).

References

Atun F (2012) Enhancing resilience of London by learning from experiences. TEMA J Land Use Mobil Environ 2:147–158

Banks JA (1968) Population change and the Victorian city. JSTOR Vic City Stud 11(3):277–289

Butler T (2007) Re-urbanizing London docklands: gentrification, suburbanization or new urbanism? Int J Urban Reg Res 31(4):759–781

Brownill S (1990) Developing London's docklands: another great planning disaster?. Paul Chapman Publishing, London

Champion AG (1987) Momentous revival in London's population. Town Country Plann 56:80–82

Chandler T (1987) Four thousand years of urban growth: a historical census. Edwin Mellen Press, Lampeter

Correll MR, Lillydahl JH, Singell LD (1978) The effects of greenbelts on residential property values: some findings on the political economy of open space. JSTOR Land Econ 54(2):207–217

Dix M (2002) The central London congestion charging scheme—from conception to implementation. imprint-Europe implementing reform in transport effective of research on pricing in Europe. An European Commission funded Thematic Network (2001–2004). Available at http://www.imprint-eu.org/public/papers/imprint_dix.pdf

Forshaw JH, Abercrombie P (1943) County of London plan. Macmillan and Co. Limited, London

Foster J (1999) Docklands: cultures in conflict, worlds in collision. UCL Press, London

Gilbert S, Horner R (1985) The Thames barrier. Thomas Telford, London

Hall P (2002) Cities of tomorrow, 3rd edn. Blackwell, London

Hall P (1989) London 2001. Unwin Hayman, London

Hall P (1988) Cities of tomorrow: an intellectual history of urban planning and design in the twentieth century. Blackwell, London

Holford I (1976) British weather disasters. David and Charles, Newton Abbot

Jonkman SN, Kelman I (2005) Deaths during the 1953 north sea storm surge. In: Proceedings of the solutions to coastal disasters conference, American Society for Civil Engineers, Charleston, South Carolina, 8–11 May 2005, pp 749–758

Osborn FJ (1950) Sir Ebenezer Howard: the evolution of his ideas. JSTOR Town Plann Rev 21(3):221–235

Owen D (1982) The government of Victorian London 1855–1889: the metropolitan board of works, the vestries and the city cooperation. The Belknap Press of Harvard University Press, Cambridge

Parker DJ (1999) Disaster response in London: a case of learning constrained by history and experience. In: Mitchell JK (ed) Crucibles of hazard: mega-cities and disasters in transition. United Nations University Press, New York

The UK Registrar General Census (1961) England and whales preliminary report (Table 6). HMSO, London, p 75

Transport for London (2009) Annual report and statement of accounts 2008/09. http://www.tfl.gov.uk/assets/downloads/annual-report-and-statement-of-accounts-2008-09.pdf. Accessed on 27 Feb 2014

Withington J (2010) London's disasters. History Press, Gloucestershire

Chapter 3
Current Flood Risk and Transportation System

Abstract This chapter sets the current flood risk situation and the pattern of the transportation system. Besides, it includes the present flood risk and the flood risk management including both structural and non-structural tools.

Keywords Flood hazard · Flood risk management · London · Transportation system

3.1 Present Flood Hazard

Approximately 15 % of all the properties in London are located in the floodplain—that is just over half a million properties where one million people live (Environment Agency (EA) no date). Approximately 70 % of the floodplain are at risk from tidal flooding, 29 % are at risk from fluvial flooding, 1 % is at risk from both (EA no date). As the tides in the Thames Estuary rise 60 cm every 100 year, the probability of flooding in London is increasing. There are number of indicated issues by the Environment Agency (no date) to explain this fact. These are:

- "The weather becoming stormier.
- The south eastern corner of the British Isles tilting downwards.
- Sea level rise.
- London settling into its clay bed".[1]

The worst scenario would be a high surge occurring at the same time with a "spring tide",[2] which occurs twice a month. Surge tides occur when low pressure reach British Isles by increasing the sea level and creating "hump" of water, which

---

[1] Environment Agency, the booklet of the Thames Barrier and associated tidal defences.
[2] Spring tide: Either of the two tides that occur at or just after new moon and full moon when the tide-generating force of the sun acts in the same direction as that of the moon, reinforcing it and causing the greatest rise and fall in tidal level (Collins English Dictionary).

F. Atun, *Improving Societal Resilience to Disasters*, PoliMI SpringerBriefs, DOI: 10.1007/978-3-319-04654-9_3, © The Author(s) 2014

moves the low pressure. Dangerous conditions would be faced if this low depression goes through Scotland into the North Sea. Besides, strong northerly winds could increase the severity of the situation.[3]

3.2 Present Flood Risk

Although flood hazard probability is increasing, because of climate change and sea level rise, the main reason of increasing flood risk in London is the *post-defence development* (Parker 1995, p. 341) and increased ownership of goods and property in the floodplain (Green and Penning-Rowsell 1989, cited in Parker 1995, p. 342). The post-defence development (Fig. 3.1) after the construction of the Thames Barrier in the 1980s, such as increasing number of population, buildings, companies and firms, and extended infrastructure in the floodplain, led to increase exposure to hazards. Moreover, more businesses have been established and more infrastructures have been constructed in the area. However, not only direct damages but also indirect damages may increase due to growing number of businesses, infrastructure and demand on traffic in the area (Parker 1995, p. 342). Having a 1 in 1000 year flooding could affect the area tremendously and lead to high consequences in terms of human causality and economic lost. As the area is the heart of the finance and business in the UK, the effects could be not only economic but also financial as well.

In Canary Wharf, the Docks were removed and the old fabric of the area was replaced with *high quality prestige type developments* (Parker et al. 1995b, p. 13) (Fig. 3.2). Moreover, the city airport was built as a part of the regeneration project right next to the old Docks (Fig. 3.3). Replacement of the existing fabric with higher quality buildings and infrastructures by increasing the density of the development has led to increase both the value of the area and the economic cost of a potential disaster accordingly. The residential and commuter population have expanded together with the increasing number of jobs in the hazard prone area.

After the 1980s by the changing economic policy, London attracted millions of people ones more in its history. London's population has increased steadily since the late 1980s and it is projected to continue increasing (TfL 2011, p. 127). As the industrialization has changed its shape and financial sector has gained importance more than manufacturing, the industrial area and the Docks became vacant. The coming population settled in these old industrial new residential locations especially in the London Boroughs of Newham, Tower Hamlet, Lewisham and Southwark. For example, the population of Tower Hamlet was 140,000 in 1981, and this number increased to 234,828 people in 2009. Between 2001 and 2009 the resident population of Greater London had increased by 5.9 % or more than 430,000 people (TfL 2011, p. 15).

---

[3] Environment Agency (EA no date), the booklet of the Thames Barrier and associated tidal defences.

Fig. 3.1 Post defence
development (Author's
archive, December 2011)

Fig. 3.2 High quality
prestige type development
(Author's archive, December
2011)

Fig. 3.3 London City
Airport (Author's archive,
December 2011)

Redevelopment of the deprived areas, especially redevelopment of the Docklands, paved the way for other problems that are income inequality and social polarization, which can be seen in all megacities. London has experienced this problem since the 1970s. The poorer population had grown in numbers, with no sign of getting better off (Parker 1999, p. 201). These tendencies affected especially racial and ethnic minorities in an imbalanced way, as they intensely concentrated particularly in the inner London. In connection with the declining value of state unemployment benefits and other social support programs because of relative income changes, higher unemployment rates greatly led to the worsening of relative deprivation (Parker 1999, p. 201).

Economic activities have transformed from manufacturing to business and financial services, construction, retail, tourism and leisure activities. The areas of the economic transformation have overlapped with the areas, which has experienced the increasing number of job growth, and consequently, the rapid population growth (GLA 2007, cited in the London Plan 2008, p. 29).

The socio-economic drawbacks mainly intensified in three adjacent inner city boroughs: Islington, Hackney and Tower Hamlets. These areas, which have exceeding number of population, predominant rental housing and severe unemployment rates, have experienced the process of gentrification that enlarges the income gaps between the rich and the poor (Gordon and Sassen 1992, cited in Parker 1999, p. 201). The transformation in the economic activities has favoured educationally qualified white workers of both sexes at the expense of men from established working-class communities (Gordon and Sassen 1992, cited in Parker 1999, p. 201). On the other hand, low-wage service jobs that require little qualification also showed an increase in number, because of flexible jobs augmentation within a growing informal sector.

3.3 Flood Risk Management

Flood risk management includes two parts: identification of present flood hazard and providing structural and non-structural defence tools to mitigate the present risk. Hazard maps, which should also include multiple and indirect hazards, sea level rise and climate change, provide knowledge on potential flooding. Regarding to **structural** defence measures, there are two types of structural defences: hard defences constructed by metal, concrete etc., and natural processes that are to provide space for floodwater. The first one involves embankments, walls and barriers. The latter includes wetlands and mudflats. Non-structural measures include the flood forecasting, warning and response systems.

In London, there are both structural and non-structural measures working like an entire system for achieving the best effect with limited resources. The flood defences along the River Thames are classified in three groups: upstream defences, interim defences and downstream defences. Embankments, Thames and Barking Barriers are the examples to structural defences in London. The non-structural tool is the emergency management system, which is highly developed and strongly hierarchical. Emergency management system includes flood forecasting and monitoring, early warning and response processes.

3.3.1 Structural Tools

In London there are six barriers including the Thames Barrier. All the structural measures including barriers and embankments are working as a whole system and designed to protect London until 2030 against 1 in 1000-flood event (Lavery and Donovan 2005). However, when the construction of the Thames Barrier, which protects central London against tidal flood, was started in 1974, climate change and sea level rise were not phenomena that were of concern, so, this could affect the life of the barrier and shorten it (Parker 1995). The Thames Estuary 2100 project (2012) established to consider the ever-changing situations and provide solutions for the next 30, 50 and 80 years (TE 2100 2012).

3.3.2 Non-structural Tools: Flood Forecasting, Warning and Response Systems

Flood forecasting, warning and response systems (FFWRS) in the UK, consist of various organizations; such as

- "Meteorological and flood defence agencies, the police, local authorities, emergency planners, and other emergency services,
- Industry, commerce, voluntary services,
- Public facilities, like hospitals, schools, rest centres, and the residents and users of the flood plains" (Parker et al. 1995a, p. 33).

As there are different actions that must be taken such as flood detection, forecasting, warning, dissemination of warnings and response to flooding, a large number of public and private organizations are involved in these processes (Fig. 3.4). All these organizations are connected to each other by communication links. If the number of organizations increases, the number of communication links increases proportionally. If a flaw occurs in one of the communication link, successive failures may spread in the entire system. Therefore, FFWRS is a system, which can be unreliable and fail easily, due to the high level of interconnectedness within the system; on the other hand, the success of the previous steps, such as enhanced flood forecasting could help to increase flood warning lead time (Parker et al. 1995a, p. 32), and consequently, it raises the quality of response and success rate of an evacuation, if it is needed.

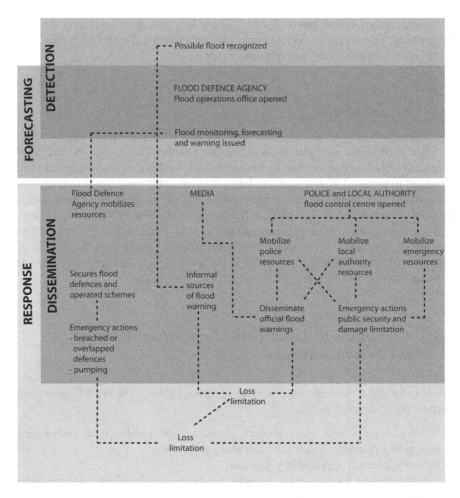

Fig. 3.4 Process of forecasting, warning and response (Parker et al. 1995a)

Parker et al. (1995a, p. 36) state that flood forecasting methods vary in different regions of England. However, generally flood forecasting is more developed for fluvial flooding than tidal ones; therefore, it is necessary to improve detection and modelling of tidal floods (Parker et al. 1995a).

During dissemination of warning and response phases, as the actions are taken locally, local authorities need to be involved in these activities. The power is given to local authorities by "Civil Protection in Peacetime Act 1986" and the "Local Government Act 1989" (section 156) (Parker et al. 1995a, p. 37). However, the main problem is that local authorities are not forced to take actions and they have the right to prioritize their funding. They may give the priority to other issues, such as health care, education, by neglecting prevention for flooding even though they are located in a floodplain (Parker et al. 1995a, p. 37).

3.3.3 Non-structural Tools: Emergency Management

Emergency Management in London is based on several plans prepared mainly by the London Resilience Team[4] at different scales and for different purposes. Current plans can be grouped in regional strategic frameworks, area specific emergency plans, thematic plans, specific organizational plans, multi-agency flood plan and other plans. The Multi-Agency Flood Plan provides guidelines for all the agencies that could be involved in a flood incident, so that they understand their responsibilities and their role during an emergency management process. Emergency management in London operates in three major groups: Gold (strategic), Silver (tactical) and Bronze (operational). These titles qualify functions, which are role related, and each title forms one of the three emergency services.

- *Gold (strategic)* Gold is the commander in charge of all other services. It has the responsibility for formulating the strategy for the incident (LR 2012).
- *Silver (tactical)* Silver attend the scene, take charge and be responsible for formulating the tactics to be adopted by their service to achieve the strategy set by gold (LR 2012).
- *Bronze (operational)* Bronze will control and deploy the resources of their respective service within a geographical sector or specific role and implement the tactics defined by Silver (LR 2012).

The Gold coordinating group (GCG) A Senior Police Officer chairs the Gold coordinating group (GCG). GCG consists of cells of various organizations which are led by the gold representatives of each cell, such as fire, ambulance, military, police, media, Environment Agency, transport, utilities, national health office etc. London Local Authority, Science and Technical Advice and Government Liaison Team are examples to these cells. In an emergency, the GCD locates at the strategic coordination centre and decides the strategic aims, objectives, and priorities in response to emergencies. This group is also responsible for the first respond and recovery (LR 2010).

London Local Authority Gold Operating Procedures The function of the London Local Authority Gold is to manage the collective response of London's local authorities to an incident. Local Authority Gold through the London Local Authority Co-ordination Centre (LLACC) directs the co-ordination of London's 33 local authorities. London Local Authority Gold gives early consideration to the wider recovery management issues, and, when appropriate, recommends to the GCG Chair that a recovery management cell would be established in accordance with the principles set out in London Recovery Management Protocol (LR 2010).

---

[4] London Resilience team facilitates the work of the London Resilience Partnership, which is a partnership of more than 170 public and private sector organisations. In an emergency, the team coordinates information from partners to maintain situational awareness of the impacts on London of the incident. Another duty of the team is to provide strategic advice on London's plans to the Mayor of London and other senior decision makers.

Science and Technical Advice Cell The Science and Technical Advice Cell (STAC) can be requested based on the information coming from the gold coordinating group chair, or the gold representative from the health protection agency, or the representative of National Health Service London, or Environment Agency. The STAC would be expected to provide advice on issues such as the impact on population health, public safety, environmental protection, food and water safety and sampling and monitoring of any contaminants (LR 2010).

Government Liaison Team The role of the government liaison team is to provide the link between the gold coordinating group, the central government overview, and response provided at the Cabinet Office Briefing Room (COBR) (LR 2010).

3.4 Present Transportation System

According to the report of TfL published in 2011, the demand for travel has increased 8 % in 2009/2010 with respect to 2000/2001, as a result of 7.1 % population and 5.5 % occupation increments. Moreover, mode of transportation in London has shifted from car to sustainable public transport, such as walking and cycling (5 % points at the trip level), which makes about 1 million trips. Furthermore, bus kilometres have increased with 32 % and underground kilometres with 9 % in 2009/2010 as compared to 2000/2001, because of the new DLR system in the Docklands (TfL 2011, p. 1).

Regarding the road traffic volumes, there were 6 % less vehicle kilometres in London in 2009 than in 2000. Volumes of road traffic in London decreased by 3 % between 2008 and 2009, reduced by 2 % between 2007 and 2008 and by 1.4 % between 2000 and 2007. The main reason of the steady decrease between 2008 and 2009 was economic recession in 2008 (TfL 2011, pp. 2–8). Consequently, there had been an increasing shift towards public transportation modes, especially within central and inner London (TfL 2011, pp. 2–8).

3.4.1 Road Traffic and Car Ownership

In 1993, the main mode of transportation was car trips (both passenger and drivers) with 10.2 million per day. In 1999, the number of car trips reached a peak of 10.5 million per day by increasing gradually since 1993. By 2000 the number of car trips had started declining gradually, and it was 9.8 million trips per day in 2009 (TfL 2011, p. 37). In addition to residential population, commuters outside London, long-term visitors and tourists constitute "daytime population" of London. According to the TfL's report (TfL 2011, p. 39), the daytime population had increased by 14 % between 1993 and 2009. Additionally, the number of trips per day had risen by 17 % since 1993.

The total road traffic for all motor vehicles decreased by 3 % within the borders of Greater London in 2009. Since the beginning of the twenty first century, there is a tendency of decline in the demand for road traffic in the central London more than the outer London, the decline had even increased between 2008 and 2009 (TfL 2011, p. 142).

In 2009, the number of trips, including trips of residents, commuters, visitors and tourists, made to, from, or within London was 24.4 million per day (TfL 2011). Between 1993 and 2007, the number of trips had increased steadily. However, the number of trips fell slightly in 2005. The terrorist attack in 2005 can be showed as a reason for this decline.

According to the National Travel Survey (2008/2009), 43 % of households in London do not have a car and 17 % of households have two or more cars. Car ownership levels have not changed significantly during the 5 years of the survey (measured by TfL's LTDS survey) (2005/2006 to 2009/2010). In Greater London, the percentage of households without a car has remained constant at 42 %. The car ownership pattern is different in the Inner and Outer London. The amount of people who own one or more cars is higher in the Outer London. Indeed, the main factor affecting car ownership is the household income. But in addition, there are other reasons defined by the report of TfL (2011, p. 69), such as availability and quality of alternative means of travel (public transport, walking or cycling), need to own and use a car (depending on the household size, structure, employment and location), lifestyle, tastes and preferences. The report states also that in London, car ownership level is stable, but car use is declining, which could be strongly related to the transport development policies and the congestion charges.

3.4.2 Congestion Hotspot

The congestion hotspots were identified in an average hour during the weekday the AM peak, inter peak and PM peak periods (TfL 2011, pp. 101–104). Congestion hotspots are spread across London with higher concentration at the central and inner London. Congestion during the PM peak is higher than the AM and the inter peaks. The hotspots in the PM peak spread mainly across the outer London. The number of the hotspots during the inter peak are less than the AM peak, however, the number of congestion hotspots in the central London during the inter peak are more than the AM peak. Congestion starts during mid-day and afternoon periods. During the weekend, the pattern of congestion is different than in the weekdays. At Saturday midday, differently from weekdays, congestion hotspots spread across much of the Greater London, while congestion in the central London is relatively low. At Saturday PM there are more congestion hotspots across London than during the inter peak period (TfL 2011, pp. 101–103).

Regarding the shift in the modes of transportation, the biggest changes occurred in car travellers and cyclers. By introducing congestion charge in 2003, the number of cars entering into the city of London has decreased steadily. In 2009, it declined again by 0.4 % compared to 2008. The increasing amount of congestion charge is the reason of this decline (TfL 2011, p. 60). Moreover, the largest percentage change is seen in cycling. The increase was by 27 % between 2007 and 2008, by 15 % between 2008 and 2009 (TfL 2011, pp. 60–61). Increasing safety, number of bicycle parking facilities, and introduction of cycle lanes and paths are the reasons of this steady increase on demand. In terms of public transport, in total 90 % of travellers use public transportation modes to enter the city during the AM peak. This share has increased by 84 % since 2000. However, all public transportation modes decreased by 4 % by 2008, while the number of rail passengers fell by 4 %, underground and DLR passengers dropped by 6 %, the number of bus passengers raised by 1 % (TfL 2011, p. 60).

3.4.3 Public Transportation

One of the advantages of today's London is well-developed and evenly distributed public transportation network; the opposite of the case in the nineteenth century. Today not only the central London but also town centres around London are accessible by public transportation modes. Highly accessible public transportation network of the city is composed of the underground, the over-ground, DLR, railways and the extensive bus network (The London Plan 2008, p. 56).

There are also several new transportation projects in connection with the development projects, and it is aimed at achieving a number of goals. First, the location of development areas has been chosen to increase sustainable transportation modes, such as walking and cycling. Second, the main concern is to choose development areas that could reduce the need to travel. As a third option development areas can be located around the key interchanges such as major rail or underground stations so as to maximize the use of the already existing transportation network (The London Plan 2008, p. 126). According to these criteria, one of the development areas is the "Thames Gateway," which necessitates more river crossings in the East London to achieve regeneration and development. In this area, the priority is given to improve access for people, goods and services between the north and south of the River Thames in the regeneration and development of the Thames Gateway region (The London Plan 2008, p. 126).

References

EA, Environment Agency (no date) The booklet of the Thames Barrier and associated tidal defences
GLA Economics (2007) Working paper 20. Employment Projections for London by sector and borough
Gordon I, Sassen S (1992) Restructuring the urban labor markets. In: Fainstein SS, Gordon I, Harloe M (eds) Divided cities: New York and London in the contemporary world. Blackwell, London

Green CH, Penning Rowsell EC (1989) Flooding and the quantification of "intangibles". J Inst Water Environ Manage 3(1):27–30

Lavery S, Donovan B (2005) Flood risk management in the Thames Estuary looking ahead 100 years. Phil Trans R Soc A 363(1831):1455–1474

LR (2012) London Resilience: command, control and information sharing protocol. London Resilience Team, London

LR (2010) London Resilience: London strategic emergency plan. London Resilience Team, London

Parker DJ (1999) Disaster response in London: a case of learning constrained by history and experience. In: Mitchell JK (ed) Crucibles of hazard: mega-cities and disasters in transition. United Nations University Press, New York

Parker DJ (1995) Floodplain development policy in England and Whales. Appl Geogr 15(4):341–363

Parker DJ, Fordham M, Tunstall S, Ketteridge AM (1995a) Flood warning systems under stress in the United Kingdom. Disaster Prev Manage 4(3):32–42

Parker DJ, Fordham M, Portou J, Tapsell S (1995b) The flood risk to London a preliminary scoping study. Flood Hazard Research Centre Publication Number 259

TE 2100 (2012) Thames Estuary 2100. Managing flood risk through London and Thames Estuary, Environment Agency. http://a0768b4a8a31e106d8b0-50dc802554eb38a24458b98ff72d550b.r19.cf3.rackcdn.com/LIT7540_43858f.pdf. Accessed 27 Feb 2014

TfL Transport for London (2011) Annual report and statement of accounts 2010/2011. http://www.tfl.gov.uk/assets/downloads/corporate/tfl-annual-report-2010-11-final-interactive.pdf

The London Plan (2008) The London Plan: spatial development strategy for greater London. London. http://www.london.gov.uk/thelondonplan/docs/londonplan08.pdf

Chapter 4
Systemic and Social Vulnerability to Flooding

Abstract This chapter starts with critiques on the present flood risk management. The chapter proceeds with the current systemic and social vulnerabilities by focusing on the downstream of the River Thames.

Keywords Systemic vulnerability · Social vulnerability · Transportation system · London

4.1 Critiques on the Present Flood Risk Management

The maps provided by the Environment Agency are the main sources to understand the present flood hazard in London. However, they do not include indirect effects of a potential flood, such as fire, disruption to transportation and traffic congestion. If a storm surge is forecasted 4 h in advance, it would create an enormous traffic congestion in the central London, because of people leaving their work and trying to reach home or outside London. Moreover, the plans do consider neither climate change nor sea level rise. The main problem of the boroughs about the present flood risk is that of having different models to define the floodplain at the local level. Developers use a model that does not identify their area as the floodplain and ask the permission to the borough for having new establishment which fits to their model. There is a need to prepare new flood scenarios by considering climate change, sea level rise, and possible indirect effects of a potential hazardous event, and to decide one single model on its feasibility and adaptability that shall be used by everyone.

Structural defence measures constitute the major part of the flood risk management system in London and generate a bundle of problems. Structural measures have been improved in the course of the time up to the present condition. However, they have several fallacies. First, the existing system was built without considering climate change and sea level rise. Second, structural defences are owned by private or public bodies, the latter includes several boroughs and Environment Agency. Maintenance of the structural defences is costly, and it would be unaffordable for private bodies to keep the maintenance level of the defences in condition "3" ("1" signifies the best condition, "5" is the worst). Third, defences are dependent on other systems such as gas or electricity; they might fail in case of a long lasting event, even though they have a back-up system, such as electricity generators.

Flood forecasting, warning and response systems work together by involving a range of public and private organizations. The system operates at different scales, changing from national to local, and contains three delicate issues needing special care.

First of all, the system is constituted by various organizations and supporting them with efficient communication links is very crucial in order to work in harmony before and during an event. Failures in communication may lead to paralysis of the entire system or misunderstanding the shared information.

The second issue is the public involvement. The end-user of the FFWRS (Flood forecasting, warning and response systems) is the public. The necessary information on forecasting and monitoring are available online, and it is possible to follow the flood warnings in the entire country at the Environment Agency's website. However, the problem is that the majority of people are not really informed and they do not register to the flood warning messaging service.

The third issue, which actually links the first to the second, is that the FFWRS functions at different scales while it is implemented mainly at the local scale. The information comes from upper levels and is conveyed to public at the local level where the existing resources of boroughs are not equal. Some boroughs are relatively well prepared. They have already informed their residents about the current flood risk and distributed flood packs that includes a torch, radio, a camera, water sanitation tablets etc. They do have a system in the main transportation nodes to inform public immediately in case of an emergency. They have improved their back up services, built flood resistant control rooms to take under control the situation in case of an emergency. On the opposite, various boroughs do not have a system to warn people. All they have is an emergency plan, as it is obligatory. However, the quality and the details of such plans vary according to the budget defined for this activity by each borough (Table 4.1).

4.2 Systemic Vulnerability: Defining the Inter-dependency

In Table 4.2, systemic vulnerabilities have been analysed in four sub-systems: urban fabric-physical structure, urban fabric-infrastructure, economic and social. The first sub-system consists of critical facilities, such as hospitals. It is crucial to know the number of beds in a hospital, its capacity and range of these services. The proximity of a hospital to hazardous areas is vital to assess the systemic vulnerability. In Fig. 4.1, there is one hospital directly located in the flood zone in the London Borough of Newham. There is another hospital in the London Borough of Tower Hamlet that is in the vicinity and out of the first level flood risk zone, and can be used for evacuation purposes in case of an emergency.

Table 4.1 Flood hazard identification for London

Flood hazard identification for London metropolitan area

	Aspect	Tools to assess		Explanation
Present flood risk	Hazard maps in different scales (containing depth, frequency and duration of flood)	Yes or no, and quality	Environment Agency provide the 1 in 1,000 year flood hazard map for the Thames Estuary	Boroughs have the responsibility to detail the existing hazard map prepared in the upper scale. The details of the map depend on the budget of the borough
	Consideration of indirect hazards	Yes or no, and quality	No! Indirect hazards are not considered in hazard maps	
	Consideration of climate change	Yes or no	No, climate change is not considered in the present flood hazard maps	The need to consider climate change is known and it will be considered in the future projects
Monitoring flood	Monitoring network	Yes or no	Yes	
	How does it work?	Qualitative	It consists of several steps in different levels; consequently, it consists of various organizations and public	Communication is very crucial. Having failures could create cascade failures in the following steps
	Quality and distribution of monitoring network	Binary		
Forecasting flood risk	Forecasting system	Yes or no and capability	Yes	

(continued)

Table 4.1 (continued)

Flood hazard identification for London metropolitan area

	Aspect	Tools to assess		Explanation
Flood warning	Does a warning system exist?	Yes or no, and quality	Yes	
	Do authorities and institutions know their duties clearly?	Yes or no, and quality	Yes	
	Do the community know what to do as a response to flood warning?	Yes or no, and quality	There is no information about this issue for London. There are studies in the smaller towns	According to the survey results, majority of the community does not know what to do as a response to flood warning
	The coverage of the system	Quantitative	The system covers all the people who have internet access and a cell phone	Registering to the system is not obligatory for individuals. So, the number of registered people is limited

(continued)

Table 4.1 (continued)

Flood hazard identification for London metropolitan area

	Aspect	Tools to assess		Explanation
Existing structural defence measures	Do they exist? What are they?	Qualitative	Walls and embankments along the Thames and other barriers, such as Barking	
	Defence standard	Quantitative	Defence standards for London is 1 in 1,000 year flood event	
	Do protection standard take climate change into consideration?	Yes or no, and quality	No	Climate change and sea level rise were not taken into consideration. It is planned to consider them during the maintenance of the defences after 2030
	Condition of defence – maintenance	Quantitative	They have been accepted in level 3. However, there is no detailed study in that	It would not be affordable to keep maintenance, as some of the defences are owned privately
	Dependency of the defence structure to other systems (such as electricity, gas, etc.)	Binary	They are dependent to electricity. Defences have back up system in case of an emergency, which won't be sufficient for long term	
	Does flood retention area exist?	Yes or no, and quality	The answer varies according to the location of the event, and according to the implementation of boroughs in the local level	

The methodology of the table adopted from the matrices of ENSURE project and the source Parker et al. (1987)

Table 4.2 Precondition of systemic vulnerabilities for flood hazard risk in London

Precondition of *systemic vulnerabilities* for flood hazard risk in London metropolitan area

	Feature	Variable	Explanation
Urban fabric physical	Public facilities	Existence Public Buildings Hospitals (bed number) Location of these facilities (proximity to hazardous area) Capacity and range of service of these facilities	There are hospitals within and in the vicinity of the flood zone
	Resources face with emergencies	Firefighting equipment (number and capacity) Emergency equipment (define them and their capacity)	Boroughs do not have sufficient emergency equipment, such as firefighting equipment, vehicles, sandbags, water pump etc. for a 1 in 1,000-year flood event
Urban fabric infrastructure	Access to critical and public facilities (hospitals, emergency rooms, fire brigades)	Road network reduction Road characteristics Expected travel time Existing transportation modes to reach those facilities and their capacity (such as helicopter, heavy machines) Existing personnel	There is no study on emergency road networks, how to access to critical facilities and expected travel time Existing personnel would not be sufficient
	Accessibility to vulnerable areas	Road network reduction Road characteristics Expected travel time	There is no study on accessibility
	Industrial areas and plants	Existing industrial areas, plants Contingency plan for Na-Tech Accessibility to plant	No data

(continued)

Table 4.2 (continued)

Precondition of *systemic vulnerabilities* for flood hazard risk in London metropolitan area

	Feature	Variable	Explanation
Economy	Assessment of the systemic vulnerability depending on economic and financial structure	Business continuity plan Access to markets Degree of dependence of production sites from lifelines Transferability to other production sites Capacity to run economy and respond crises Capacity to invest in recovery and take prevention action	The importance is given to the business continuity plan in every interviewed Borough. The emergency department of boroughs provide guidance to businesses for preparing business continuity plan. Businesses have facilities also outside the central London ready to be used in case of any kind of emergency
Social	Community preparedness in case of event	Established protocols for use of resources Overlapping responsibilities among agencies Training Rehearsals Availability of temporary shelters Availability of temporary location Guidance for the foreign aid teams	Training in the tactical level is rather effective. Rest centres of every borough are ready. Emergency personnel have the list of vulnerable population and update the list every 6 months. Agencies know their responsibilities. However there is no interaction among neighbour boroughs
	Individual preparedness in case of event	Access to understandable information	Information provided to public is accessible and understandable
		Trust in information provides	Trust to information providers is very high
		Existence of individual contingency plan	According to the result of the survey, individuals do not have contingency plan

The methodology of the table adopted from the matrices of ENSURE project and the source Parker et al. (1987)

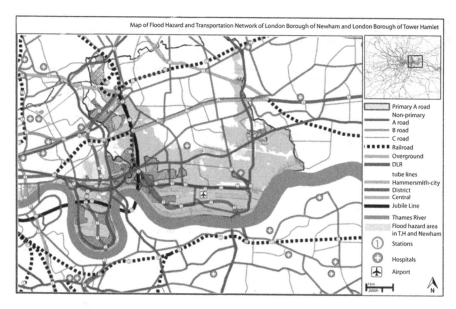

Fig. 4.1 London borough of Newham and London Borough of Tower Hamlet: Map of flood hazard, transportation network, and hospitals

In case of an emergency, boroughs do not have sufficient number of personnel and resources, such as firefighting equipment, vehicles, sandbags, water pump etc. to act independently in case of a 1 in 1,000-year flood event. In such a situation, external help will be needed from other boroughs, Environment Agency, Metropolitan Police and London Fire Brigade. If one considers that in a major event all the boroughs will face scarcity of resources, help would come from other cities in the vicinity of London.

The transportation system is strongly dependent on other systems, such as communication and electricity. In terms of providing an efficient mobility, there is no study for London about potential emergency road networks, how to access and expected travel time to critical facilities.

The importance is given to the business continuity plan in every interviewed Borough. The emergency department of boroughs provide guidance to businesses for preparing the business continuity plan. Businesses have facilities also outside the central London ready to be used in case of any kind of emergency.

Although the number of staff is limited, training of the staff is rather effective, so this increases the chance of an effective response during emergencies. Additionally, rest centres of every borough are ready. Emergency officials in each borough have the list of vulnerable population and update the list every 6 months. They know their responsibilities, however, there is no interaction among neighbour boroughs in terms of preparing emergency plans.

4.3 Social Vulnerability: Describing Possible Behaviour Patterns for the Users of Transportation System

To produce primary data and to have some insights on the social vulnerability, three surveys have been conducted. The first one was conducted with employees of the emergency management system, so as to understand their response capability to disasters. The second survey was conducted via a questionnaire consisting of 11 questions asked to the transportation system officials. The questionnaire includes questions on emergencies that transportation employees are dealing with, the procedures they adopt during emergencies, and their opinion about public response to emergencies. The third questionnaire was conducted with the lay people at two different locations by aiming to understand individual's and community's awareness of risk in these areas.

It was decided to conduct the survey with public after being informed about the flood hazard information and warning programmes. The purpose of the survey was to question the awareness of public on risks and information programmes. According to the survey results, awareness of public is rather low. Table 4.3 gives a general idea on the result of the survey. The detailed results of the surveys are given in the following parts.

4.4 Focusing on the Downstream of the Thames River

In London, there are 33 boroughs in total; 17 of which are along the River Thames. The case study area covers the upper Thames, which includes Newham, Greenwich, Tower Hamlet, Southwark and Lewisham Boroughs of London (Fig. 4.2). In this area, it is possible to track the economic and demographic changes in London since the early 19th century. By industrialization, corresponding the need to stock products, a dock system has been impounded. Later, the Docks became idle due to changes in the technology.

Application of the methodology

- Understanding the present flood risk
- Understanding the risk of flooding to the transportation system
- Systemic failure: defining dependency of the transportation system on other subsystems.
- Social Vulnerability: describing possible behaviour patterns for users of the transportation system.

4.5 Transportation System Under the Risk of Flooding

Transportation system consists of rail, road, air, and sea transportation modes with various elements, such as vehicles, terminals, depots and petrol filling stations etc. The elements of the transportation system can be also divided as lines and nodes.

Table 4.3 Precondition of social vulnerabilities for flood hazard risk in London

Precondition of social vulnerabilities for flood hazard risk in London metropolitan area

System	Variable	Data type	Explanation—according to the result of the survey
Social Demographic	Number of day and night population	Quantitative	The residential population of London is 7.6 million and the number of trips made to, from, or within London was 24.4 million per day
	Proximity of population to expected flooding and hazardous materials	Quantitative	In the Thames Gateway 1.5 million people already live in the region, 160.000 new homes and 225.000 jobs are to be created by 2016
	People with difficulties to comply with evacuation orders, difficulties in escaping	Quantitative	The boroughs have the list of the people who would have difficulties in moving in case of an emergency
	People with pets	Quantitative	No data
	People with and without car	Quantitative	In the inner London %57 of population do not own a car
	People without driving capabilities	Quantitative	No data
Community's awareness of risk	Participation in development, prevention and mitigation strategies	Qualitative	Population participate in the events which are related with them and their area, however, there is no awareness of risk and accordingly no participation on the prevention and mitigation related events
	Education programs	Qualitative	
	Capacity to invest mitigation	Qualitative	No data
	Access to flood information	Qualitative	Flood information is published online in the EA's website
	Training and experience of population	Qualitative	
Individual awareness of risk	Perception and awareness of risk condition	Qualitative	Majority of people do not aware risk condition
	Awareness of the education programmes	Qualitative	Majority of people do not aware of the education programmes
	Population's individual preparation	Qualitative	Majority of people do not have individual preparation
	Awareness of the ways to access to flood information	Qualitative	Majority of people do not aware the ways to access to flood information

The methodology of the table adopted from the matrices of ENSURE project and the source Parker et al. (1987)

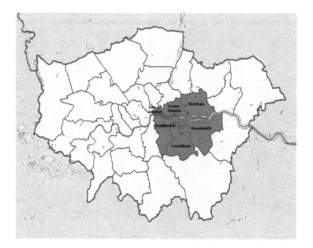

Fig. 4.2 Case Study Boroughs

Tube lines, railroad, DLR, over-ground and roads in four levels (primary "A" road, non-primary "A" road, "B" road and "C" road) are considered as the lines in the transportation network. Moreover, stations of tube lines, railroads, DLR, over-ground and airports have been defined as nodes. Among critical facilities, hospitals have been selected to be added to the edges. To improve the map other critical facilities, such as schools, public facilities, power supplies and petrol filling stations should be added as other edges (Figs. 4.2 and 4.3).

The edges are not connected at the same level. Some stations consist of multiple tube lines that make them more connected than the others. If one of these stations is flooded, more people can be affected than in the case of having a flood event in a station which includes a single line only. Some of the hospitals are well connected with the rest of the city by being located in the central locations, where there are different transportation modes. Instead, some others are located in remote places, with much less accessibility. All hospitals should have been considered, as they would need access either to evacuate patients, if the facility is in the flooded area or receives patients from the rest of the city and from evacuated hospitals.

To understand the present flood risk to transportation system, the map of the transportation system has been overlaid with 1 in 1,000-year flood hazard map, which is prepared by Environment Agency.[1] In Fig. 4.3, the result of the overlay is shown for the London Borough of Newham and Tower Hamlet. In this area, there are 28 tube and DLR stations, 4 railway stations, an airport, one hospital and one medical centre at risk of flooding. Flooded stations in the tube network, such as in this one may have repercussion in all the other connected tube networks. The map also indicates priority areas, such as hospitals that are in the middle of the

---

[1] http://www.environment-agency.gov.uk/.

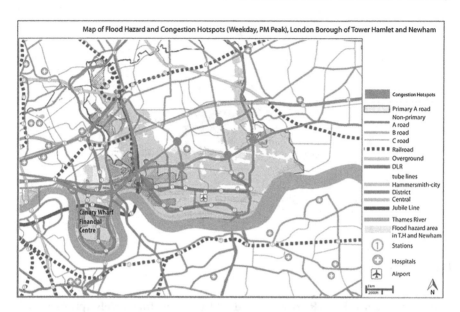

Fig. 4.3 Map of flood hazard and congestion hotspots (weekday pm peak) in London borough of Newham and Tower Hamlet

floodplain (Fig. 4.3). Besides, if this kind of overlaying is extended to the entire city, it can also help to enhance resilience of the underground network by seeing the vulnerable parts of the network and act accordingly. Moreover, Fig. 4.3 indicates the weekday PM congestion points, three of them located directly in the flood zone. In case of an evacuation, the main problem encountered in the area would be traffic congestion, which will create problems in the rest of the city as well. Not only how to distribute evacuation warning but also an effective evacuation plan shall be prepared by providing alternative routes, traffic control, controlling the number of vehicles that are already inside the zone which influence both collective and individual response to evacuation (Chen and Zhan 2008, p. 27).

References

Parker DJ, Green CH, Thompson PM (1987) Urban floods protection benefits: a project appraisal guide. Aldershot, Gower
Chen X, Zhan FB (2008) Agent-based modeling and simulation of urban evacuation: relative effectiveness of simultaneous and staged evacuation strategies. J Oper Res Soc 59:25–33
ENSURE Project—Enhancing resilience of communities and territories facing natural and na-tech hazards. EU FP7 Project. http://ensure.metid.polimi.it/web/guest/summary-ensure-project

Chapter 5
Societal Response

Abstract The fieldwork in London consists of three kinds of survey including in-depth, structured and semi-structured questionnaires, in addition to analysis of secondary data sources, such as official reports, newspapers etc. This chapter includes the methodology and the results of the three-levelled survey. The first level includes questionnaires conducted directly with the emergency personnel from different departments in several London Boroughs, Environment Agency, Port of London Authority, Thames Barrier and Transport for London. The aim was to understand the emergency protocols, flood hazard, forecasting, monitoring and early warning systems. The results of the first level helped to understand the mechanism of emergency protocols in London. The second level consists of semi-structured interviews with the employees of transportation system to understand their awareness of risk, awareness of their responsibilities during emergencies and to gather information on behaviour pattern of public during emergencies from experiences of transportation personnel. The third level includes face-to-face structured interviews with lay people to understand their perception, their preparedness level, awareness of risk and available information.

Keywords Awareness of flood warning · Awareness of information programs · Awareness of risk conditions and risk of flooding · Individual preparedness · Perception of risk conditions and risk of flooding · Transportation system

5.1 Methodology of the Fieldwork

5.1.1 Data Collection

Fieldwork consists of both analysing existed data and generating author's own data. Both primary and secondary data have been used. The latter have been collected from the reports, archives, literature and newspapers. Primary data have been collected through direct observations, interviews and surveys.

F. Atun, *Improving Societal Resilience to Disasters*, PoliMI SpringerBriefs,
DOI: 10.1007/978-3-319-04654-9_5, © The Author(s) 2014

Primary sources are given as in the following

- Fieldwork observation in London including surveys of the physical, social, economic and organizational environments has been conducted to understand both vulnerabilities and resilient capacity to response and recovery.
- In depth interviews with decision makers from governmental, private and non-profit organizations participating in different phases of disaster risk management.
- Semi-structured interviews with employees of diverse transportation modes to understand their awareness of disaster risk and capability to deal with emergencies.
- Structured questionnaires with public to understand their awareness on the existing risk, tools and information to prevent risk and their preparedness level.

Secondary sources are given as in the following

- Analysis of official reports, archival records, and literature
- Analysis of newspapers.

Regarding to the secondary data collected for the London case study:

- Most of the secondary data were collected via Internet, as all the reports are open to public in the websites of Environment Agency,[1] Met Office,[2] London Fire Brigade,[3] London Resilience Team[4] and Transport for London.[5] The general plans related with planning for emergencies in London are given in the following website: http://www.london.gov.uk/priorities/london-prepared/ preparing-london/planning-emergencies-london.
- Some of the boroughs provided the specific emergency planning and rest centres reports during the questionnaires. The others have the emergency planning reports in their websites open to public. One of the boroughs, among the interviewed ones, did not want to share the emergency planning report due to security reasons.
- Newspapers provided valuable information on the recent flood events, reaction of public to the events, and technical or organizational failures during the event.

5.1.2 Survey

The author generated primary data by conducting three types of questionnaires with the following stakeholders: decision makers, key personnel of the transportation network and lay people living in hazard prone areas. Below there are the six basic issues followed in the survey:

---

[1] Environment agency: http://www.environment-agency.gov.uk/.

[2] Meteorological Office: http://www.metoffice.gov.uk/.

[3] London Fire Brigade: http://www.london-fire.gov.uk/.

[4] London Resilience Team: http://www.london.gov.uk/priorities/london-prepared/home.

[5] Transport for London: www.tfl.gov.uk.

- To determine the related organizations that are involved in the subject, both disaster management and transportation.
- To collect reports and archival data of the related organizations.
- To set up a representative sample from public, private and non-profit organizations involved in transportation or disaster risk management for conducting questionnaires.
- To conduct questionnaires with the sample key personnel of the representative sample.
- To conduct questionnaires with the supervisors in different transportation modes in the areas prone to hazard.
- To conduct questionnaires with lay people living or working in the disaster prone area.

5.1.2.1 In-Depth Interviews Conducted with Stakeholders from the Relevant Organizations

In London in-depth interviews were conducted with the personnel of the five boroughs (London Borough of Greenwich, Lewisham, Newham, Southwark and The City of London), Environment Agency, Thames Barrier, FirstGroup Global Transport Company, Port of London Authority, National Health Service, Lewisham University Hospital and London Development Agency.

5.1.2.2 Semi-structured Questionnaires Conducted with Employees of Underground, Over-Ground, DLR and Railway

Questionnaires with the employees were conducted in three days starting from Embankment towards Eastern London stations. In total, 20 stations (Underground, over-ground, DLR, Railway) were visited, and 10 employees accepted the questionnaire request. The stations were chosen according to be prone to flooding. In the study area, there are 45 stations those prone to flooding and it was aimed to conduct questionnaires with 20 % of them, which was achieved by conducting 10 questionnaires. Besides, the questionnaires was carried out in Russell Square and King's Cross Stations, the first being prone to flooding, the latter because of the tragic fire which occurred in 1987.

Supervisors of each station answered the questionnaires. Interviews conducted in Underground (Mansion House, Cannon Street, Monument-Bank, Tower Hill, Russell Square, Kings Cross), DLR (Canary Wharf), Over-ground (Shadwell) and Railway (Limehouse).

5.1.2.3 Structured Questionnaires Conducted with People at the Public Level

In London, the interviews with people at the public level were carried out on the both sides of the River Thames. The aim was to conduct questionnaires with people from different classes, ages, genders, and backgrounds. Two areas have been chosen: Canary Wharf and the centre of Lewisham, which are different from each other in terms of development history, demographic data, economic and social structures. The interviews were conducted in the community centres so as to get a certain population mix, such as students, workers, unemployment people, people with children and elderly.

In Lewisham, questionnaires were conducted at the Lewisham Community Centre, with people who were entering the waiting lounge. It took an average of 12 min for each questionnaire depending on who was responding it. The author had been rejected 3 times and had conducted 31 valid and 1 not applicable questionnaires.

The questionnaires in Canary Wharf at London Borough of Tower Hamlet were carried out in the IDEA Community Centre. The same survey methodology, which was used in Lewisham, had been used in Canary Wharf as well. Here the author had been rejected 10 times and had conducted 30 valid questionnaires.

In Tower Hamlet, the profile of people was quite different than Lewisham. In Lewisham people were willing to do the questionnaire in the beginning, however, they lost their interest with the detailed questions and information requested. The Tower Hamlet case was quite the opposite, as people were concerned to do the questionnaire at first, but found a greater sensitivity towards the topic while answering the questions. Three people declared that thanks to the questionnaire they realized that they had no knowledge on flooding, although they dwell in the flooding zone.

5.2 First Level: Interviews with People at the Organizational Level

The aim of the interviews at the first level is to have clear understanding on the present disaster risk management system and the current disaster risk in the area. In depth interviews with the upper level officials provide an understanding of what the officials think about the disaster risk management system. During the interviews, upper level officials reported that workers in diverse transportation modes know how to behave during emergencies. Besides, they indicated that all the related materials are accessible to public online, so public can easily get the necessary information. These two points helped to construct the structure of the questionnaires with employees of the transportation system and people at the public level. Moreover, the interviewed personnel provided necessary materials, such as emergency plans and annual reports to the author. The gathered data and the results of the first level interviews are given in the related parts within the book.

5.3 Second Level: Questionnaires with Employees of the Transportation System

The questionnaire includes 11 questions regarding staff's emergency experience and the problems that they are dealing with during any kind of emergency. The aim of the questionnaire was getting more information about the behaviour pattern of people, to understand how their behaviour patterns affect the transportation system and how the transportation system officials deal with such kind of problems. Underground personnel are dealing with emergencies very frequently; however, such situations can be considered limited or minor, such as a false fire alarm, persons stuck in the stairs, falling on the rails. The survey was carried out with employees in the floodplain stations, in Russell Square and King's Cross stations. The latter two are not in the flooding zone but have experienced the former terrorist attack in 2005, the latter the King's Cross Fire in 1987.

According to the interviewed personnel, the most likely disaster in London is fire, followed by terrorism and floods (Table 5.1). The result shows that everyday routine affects how people perceive the disaster risk.

The likelihood of the underground station being flooded has been asked to the respondents. They all recognized the risk of flooding and they said that the probability of flooding is high according to various information sources. It could be due to heavy rain, fluvial or tidal flood. The underground pumps more water every single day than any other institution, including the Water Company. For example in Victoria station, there are three giant pumps behind the entrance wall. One is always under repair and the other is always under reserve. If they do not work, there would be flooding in 20 min in any day.[6]

Some officials reported that during the Second World War, there were great flooding gates' out in different parts of the tube network to avoid the flooding of the whole system. In case an inundation occurred in a part of the system, these gates were closed. Most gates are out of service now. The system could be used also today, however, information about the gates' location is not reliable whilst it would be crucial to know how one tunnel relates to the other to locate those gates.

Moreover, in transportation modes, employees are participating in drill exercises twice a year and all respondents reported that they have taken part in at least one of them. The majority of the respondents of the questionnaire said that those exercises were satisfying. Only a small percentage of the respondents were neutral to this question (Fig. 5.1).

For example, in one of the drills at the Bank station, a chemical attack was simulated. The question of the drill was: "what would happen if there was a chemical attack to a train with full of people?" Besides, they assumed that the driver was injured as well. One of the lessons learned was that sending emergency personnel in with proper clothing to start taking all people out one by one was too

---

[6] The information was attained at the interview with the former CEO of TfL.

Table 5.1 Likeliness of disasters

Disaster type	Rank
Fire	1
Terrorism	2
Flooding	3
Technological risk	4
Snow storm	5
Ash crises	6
Climate change	7
Epidemics	8
Drought	9
Earthquake	10

Fig. 5.1 Satisfaction level of drill exercises

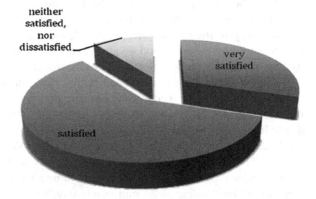

much time consuming. That led them to consider other options. As their aim was to get people out from the place of the chemical attack, they decided to use the train as a rescue vehicle. With a special little trolley that is put quickly down on the track, and the emergency personnel who knows how to operate the train, they succeeded to get the train out.[7]

Sixty percent of the underground system is above the ground and the rest of the system is actually underground. The drills were done by considering this difference and organized according to the need of different planning and experimenting conditions. Multiple hazards have been exercised during the drills. However, a multiple attacks scenario at the exact same time had never been studied until the terrorist attack in July 2005.

Furthermore, the procedure for dealing with an emergency was asked within the questionnaire as well. The respondents said that during an emergency there are two basic principles: safety of the staff and customers, and taking actions accordingly. If there is an emergency, they need to sound the alarm as soon as a dangerous

[7] The information was attained at the interview with the former CEO of TfL.

situation is identified so as to evacuate people. As previously seen in the Kings Cross fire example, being late to evacuate people could lead to high death toll.

The basic steps to follow during an emergency are:

- to inform the control centre.
- to stop people coming in.
- to announce.
- to evacuate.
- emergency people coming in.
- do not let people in until all clear.

Managing crowd is the main issue during an emergency and the most difficult one due to some issues, such as communication with public, people with disabilities and physical layout of stations. Communication is stated as one of the main problems during an emergency, because of people listening to personal stereo and language barrier. Moreover, some people ignore the alert and sometimes they overreact. People with wheelchair, or with some disabilities need close attention and help. In terms of physical deficiencies of stations, accessibility within stations and narrow platforms can be given as examples. Both of them increase the difficulty of managing crowd during an emergency at the stations. Even in the daily routine, people have difficulties of using hundreds of steps. Using crowded stairs during an emergency decreases evacuation's efficiency. The last reported problem during an emergency is the lack of staff, as the situation has to be dealt with the staff at the station present at that moment. If the situation is difficult to take under control with the present staff, the supervisor could call the London Metropolitan Police.

Respondents were asked also about specificities in case of an emergency due to flooding, only one supervisor out of ten was aware that there are different kinds of emergency protocols for each disaster types including flooding. Other officials said that they are following the fire protocol for any kind of disasters. However, handling a flood would require specific steps needed to be carried, such as understanding the source of the water, deciding to close the station partly, and extending the closure depending on the location of inundation. During any emergency, supervisor of the station decides the action and informs the controller. However, they follow the fire procedure for evacuating and warning customers. Other stations are informed about the situations and trains do not stop at the station where there is an emergency. They open all the doors and yellow punches to evacuate the customers.

Regarding to response of public in case of an emergency, one of the former CEO of the TfL said that response of public strongly depends on how much information is provided to them. If people are informed about what is occurring, they can be very tolerant. People have a tendency to protect the underground in a way as it is an icon of London. If people are not informed about what is happening, the reputation of the London Underground would get down quickly.

In general, respondents of the questionnaire agreed that people want to know what is happening. They continued saying that after the bomb attacks in 2005, people are very attentive and follow instructions quickly. Generally, they react well, but sometimes, they are argumentative, they refuse to evacuate as they have their own priorities.

5.4 Third Level: Questionnaires with People at the Public Level

Public survey was carried out on the both sides of the River Thames in Lewisham and Canary Wharf to have a sample of people from different classes, ages, genders and backgrounds. The questionnaire is made up of six parts. The first part is aimed at understanding the respondents' profile and consists of 8 questions. Detailed information on the profile of the respondents is given in Appendix C. The following five parts have 20 questions in total and these parts are:

- Part 1: Perception and awareness of risk condition.
- Part 2: Awareness of flood warning.
- Part 3: Awareness of access to information.
- Part 4: Awareness of the information programmes.
- Part 5: Population's individual preparation.

In total 61 valid questionnaires were carried out. Albeit the number of the questionnaires is not sufficient to represent the entire population in these two boroughs, the results of the questionnaires provided significant insights for understanding the present conditions in terms of perception and awareness of public on risk and existing information, and consequently social vulnerability.

5.4.1 Respondent Information

In Tower Hamlet, the profile of people was quite different than Lewisham. In Lewisham, people were willing to fill the questionnaire at the beginning, however, they lost their interest with the detailed questions. In Tower Hamlet the case was quite the opposite, people were hardly convinced to do questionnaires, but demonstrated a greater sensitivity towards the topic while conducting the questionnaire. Three people declared that thanks to the questionnaire they realized their lack of knowledge on flooding, although they dwell in a flooding zone.

Moreover, 52 % of the all respondents' properties are located in the range of a potential flood source, which is the Ravensbourne River in Lewisham, the Thames River or the Docks in Canary Wharf (Fig. 5.2). It can be indicated that the properties of half of the respondents are prone to flood hazard, which could be fluvial or tidal.

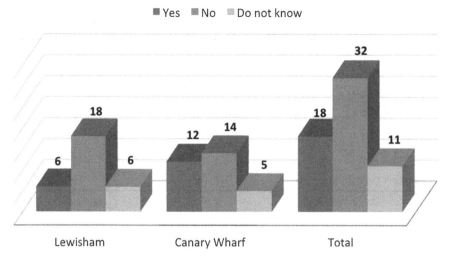

Fig. 5.2   Property in sight of potential flood source

5.4.2 Perception and Awareness of Risk Conditions and Risk of Flooding

To understand perception and awareness of risk conditions and flooding risk, the respondents have been asked to name the likelihood of ten disaster types in London (Table 5.2). In both Lewisham and Canary Wharf, terrorism and snow storm are at the first two ranks. Before conducting the survey, it was expected that ash crises could be within the first five potential disasters, as it occurred two years before the survey. As people have experienced it, they would have thought that it happened and could happen again, but the results were totally different. The respondents think that it is very unlikely that the city may be struck by an ash crise and an earthquake. So, these disasters have the ninth and the tenth ranks in the list. However, the order of the hazards differs between the third and the eighth rank (Table 5.2). Flooding is at the fifth rank in Lewisham and at the six in Canary Wharf. As in Lewisham, it has experienced flood disaster recently due to alluvial flood from the river, flooding hazard is more likely to occur according to residents in Lewisham.

Even though respondents think that their property is in sight of a flood source and there is risk of flooding in London, they do not think that their homes are at risk of flooding. Correspondingly, while 65 % of the respondents think that their homes are not at risk of flooding, 18 % think that the risk is a fair amount, and just 8 % think that their homes are at risk of flooding (Fig. 5.3). The same tendency is seen also for work and school. Moreover, 80 % of the respondents are not worried about risk of flooding and only 20 % are worried (Fig. 5.4).

Table 5.2 Perception and awareness of disasters

	Lewisham	Canary wharf	All respondents
Snow storm	2	1	1
Terrorism	1	3	2
Fire	4	2	3
Climate change	3	7	4
Technological risk	6	5	4
Flooding	5	6	6
Epidemics	7	4	7
Drought	7	8	8
Ash crisis	9	9	9
Earthquake	10	10	10

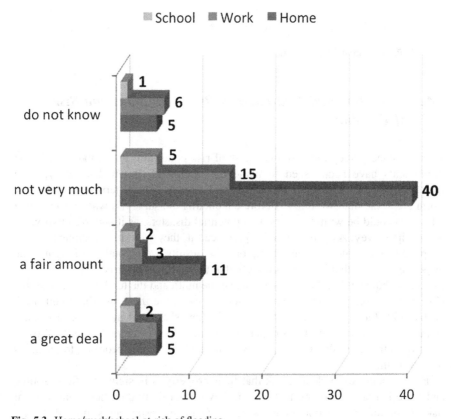

Fig. 5.3 Home/work/school at risk of flooding

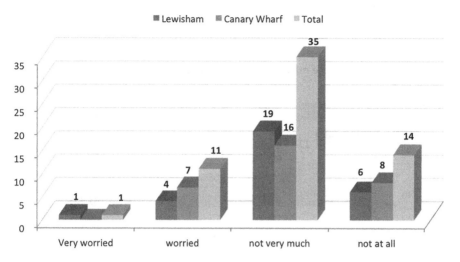

Fig. 5.4 Whether being worried about the possibility of home/work/school being flooded

5.4.3 Awareness of Flood Warning

To understand respondents' experience on flood warning, in the questionnaire they were asked whether they have ever received a flood warning or not. Only two people indicated that they received flood warning once. The person in Lewisham said that he received it when he was a child living in the county with his family. The one in Canary Wharf said that he received the warning at his workplace, which is registered to the warning messaging service. So, in total 96.7 % of the total respondents have not received a flood warning in their entire life (Fig. 5.5).

The flood warnings signs (Fig. 5.6), which could be found in the Environment Agency's website, had been showed to the respondents and they were asked whether they have ever seen them before or not. Eighty two percent of the respondents said "no" and added that they did not know the meaning of the signs as well, and 18 % answered "yes", because they saw them in the books for the driving exam. One person said that he had seen it while driving in the county, and another one said that he had seen them abroad.

Authorities' reliability in the eyes of the respondents holds importance to forecast behaviour of people when they receive a warning. If they rely on authorities, they are more likely to attend warning and get precautions. If trust in authorities is low, people tend to disregard warnings and continue what they are doing at that moment as if there is no danger. As it can be seen in Fig. 5.5, respondents in Canary Wharf rely on authorities (83 %) more than the ones in Lewisham (70 %).

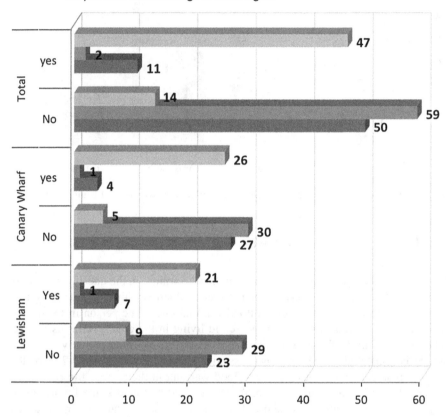

Fig. 5.5 Awareness of flood warning

Fig. 5.6 Flood warning
signs (EA)

▨ Are you aware of any information programmes on flood response / improwing conditions of home in case of flooding?

▨ Do you know the environment agency's flood warning website?

▨ Have you registered with flood warning messaging service?

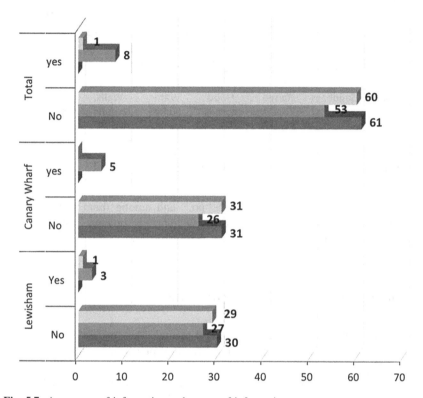

Fig. 5.7 Awareness of information and access of information

5.4.4 Awareness of Information Programs and Access of Information

The Environment agency's website provides all kind of information about flooding for the entire UK and it is possible to follow flood warnings online. For understanding awareness of respondents, they were asked whether they know the website or not. While 87 % of the respondents do not know the website, and 13 % said that they know it, however, just one respondent said he entered the website once (Fig. 5.7). Moreover, none of the respondents has registered with the messaging service (Fig. 5.7).

Another question in this part was asked to understand their awareness of information programmes on flood response, or improving conditions of home

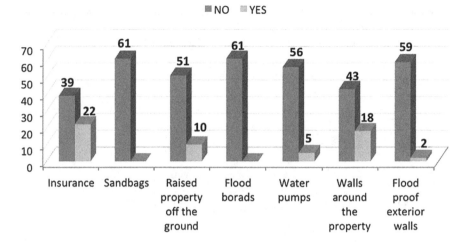

Fig. 5.8 Individual preparedness to flooding/protection

against flooding. Sixty people responded "no," and among them just one was aware of the information that is provided by the agencies, however, he never got interested or had detailed knowledge about them (Fig. 5.7).

5.4.5 Population's Individual Preparedness

The last part of the questionnaire was aimed at understanding the respondents' preparedness and the structural and/or non-structural precautions taken to protect their homes against flooding.

The list of the structural (sandbags, raised property of the ground, flood boards/gates, water pumps, walls around the property, flood proof exterior walls) and non-structural (insurance) precautions for flooding was examined by the respondents where they were also asked which of the precautions were present in their properties. In Lewisham 33 % of the total respondents' property is insured. Moreover, some respondents have walls around the property (13.3 %), dispose of water pumps (6.6 %) and raised property off the ground (6.6 %) (Fig. 5.8). In contrast to Lewisham, in Canary Wharf 38.7 % of respondents have insurance at their property, 25.8 % have raised property off the ground, 9.6 % have water pumps, 45 % have walls around the property and 6 % have flood proof exterior walls, no one has sandbags or flood boards/gates (Fig. 5.8).

The respondents have been asked whether they have undertaken any of the following actions, such as checking the level of river, the flood warning and the weather forecast (Fig. 5.9). In addition, it had been asked what is the frequency of taking these actions. As it can be expected, people check the weather forecast more often, mainly because the information about the weather is easy to find everywhere,

■ Checking the level of river   ■ Checking the flood warning   ■ Checking the weather forecast

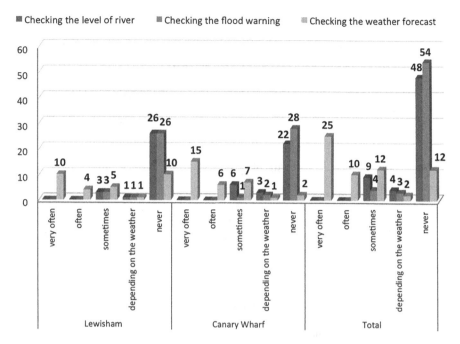

Fig. 5.9 Individual preparedness to flooding/monitoring and control

such as in newspapers, on screens in public areas, on television, in mobile phone or in internet. 88.5 % of respondents said that they had never checked flood warnings and 78.6 % said that they had never checked level of the river.

5.5 Describing Possible Behaviour Patterns for the Users of Transportation System

During an emergency people react immediately on the basis of their knowledge that they gained previously. Taking precautions is crucial before the occurrence of an event. However, as it is stated in the results, preparedness of public is generally low. The questionnaire consisted of questions to understand potential behaviour of the respondents in case of an emergency, evacuation and the degree of trust in authorities. According to the results, majority of the respondents (87 %) do not know what to do during an emergency. The other 13 % said that they would call emergency line to learn what to do (Fig. 5.10). Furthermore, while 59 respondents (97 %) said that they have never been informed about any evacuation plan by authorities, 2 people (3 %) said that they have been informed at their workplaces (Fig. 5.10).

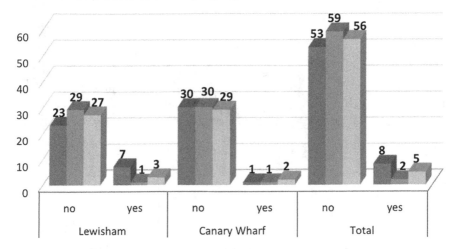

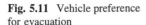

Fig. 5.10 Preparedness to emergencies

Fig. 5.11 Vehicle preference
for evacuation

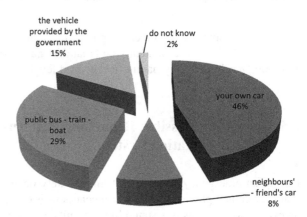

The next question asks whether they know where to evacuate to in case of an emergency. Again, the vast majority of the responses (92 %) are negative, as they said that they do not know where to evacuate to, if they need to do so. Eight percent said that they would go to the borough's main building to ask for help. Some of them said that, they would go to other family members' homes that are outside London.

The respondents have been also asked if they need to evacuate, which vehicle they will use. For this question, possible answers were grouped into five: their own car, neighbour's/friend's car, public bus/train/boat, a vehicle provided by the government, and do not know. Fourty six percent chose their own car for evacuation. Twenty nine percent said that they would use public bus/train/boat, 15 % prefered vehicle provided by the government (Fig. 5.11).

Chapter 6
Conclusion

Abstract This chapter sets the way towards resilience, and covers discussion and conclusion sub-sections. The discussion part comprises relationships among results and generalizations. Besides, new observations, new interpretations, and new insights that have resulting from the research are given in the conclusion sub-section.

Keywords Disaster resilience · London · Transportation system

6.1 Towards Resilience

There are three phases for improving resilience capacity of a community. These phases are prepare for, respond to and recover from a disaster. Although the main focus of the book is the transportation system, due to interconnectivity among systems, all the aspects that are needed to enhance resilience have been considered within the book. Accordingly, London has been studied by considering various characteristics, the phases of its development, changing vulnerability and hazard concept, structural and non-structural hazard mitigation measures (Table 6.1).

The first phase of disaster preparedness includes mitigation measures, such as limiting the exposure to hazard and diminishing direct, indirect impacts of hazard and sharing the losses. As for limiting the exposure to hazard, the development within hazard zone has to be restricted, and new development areas must be selected outside the hazard zone. Moreover, land acquisition, land use ordinances, density restriction, community relocation and having disaster management plans are the other measures to limit the exposure in the hazard zone.

In the London case, after the construction of the Thames Barrier, the floodplain area along the River Thames opened to development with the name of the Thames Gateway Project. Today, the area has been protected by the barrier and the embankments, however, during the construction of the defences climate change and sea level rise have not been considered. The exposure in the area has increased

Table 6.1 (continued)

Prepare for a disaster mitigation measures	Respond to a disaster early warning	During emergency	Recovery and reconstruction after a disaster	
			Social	Healing injured and traumatized community
				Bringing together separated families
				Identifying dead
			Institutions	Liability
				Transparency in funds
				Ability to learn from past events

tremendously due to the construction of middle and high-income class residential buildings, and the establishment of the high technology, valuable infrastructure. Moreover, the number of day and night population have increased due to the increased number of residential population, job opportunities and attraction points for visitors/tourists. Although probability of having a catastrophic flood event is low, due to increasing exposure and investment in the area, the consequences of an incident would be very high and costly.

In the same area, investment on transportation system has also increased to connect the area both to London by DLR (Docklands Light Rail) and to the rest of the world by the City Airport. Because of lack of redundancy of roads and rail networks, any disruption on the existing transportation system in the area could isolate the area from the rest of the city, although the area is well connected to London. Furthermore, infrastructure failures in case of an incident could create cascade events, such as disruption to transportation, traffic jam, disrupted connections and financial effects.

Secondly, to mitigate risk of both direct and indirect impacts, a built environment shall be reinforced to resist physical forces of tidal waves, with the construction of sea walls, leaves, fire breaks, quarantine, dams or barriers, which is well developed and organized in the study area and London in general.

The third issue, as to mitigate losses, is sharing the losses by insurances, relief funds, personal savings, financial and medical services, and indeed facilitating permanent return of residents who had to evacuate as soon as possible.

To respond to a disaster there are two aspects, early warning and emergency management. The success of an early warning system depends on the success of the procedures coming before an early warning such as forecasting, monitoring, disseminating warning etc. Monitoring of a hazard is a continuous activity, then possible hazard is forecasted and people are warned and informed about it to take the essential precautions before the occurrence of a hazard. While disseminating warning, the language of warning has to be transformed from scientific one to the one that can be understood clearly by public.

Regarding to infrastructure, first of all, the damage has to be mapped by a quick survey on the damaged parts of the network during an event. The resources and existing spare materials have to be checked before the occurrence of an event, if more material is needed, the order has to be done as soon as possible. The availability of personnel for repairing is important as much as availability of resources, because the number of personnel defines the needed time for the workload. After a disaster occurs, there would be people who need physiological care, in addition to remedies for physical injuries. The dead has to be identified and separated, families have to be gathered, as they would be far apart during the occurrence of disaster or evacuation.

Recovery after a disaster can be studied by dividing the city into physical, infrastructural, economic, social and institutional systems. To recover the physical structure, the first thing that has to be done is assessing loss and damages before starting and reconstructing the built environment. In some cases, assessment and repairing works done simultaneously to start operating the system as soon as possible. This can be possible in small and close systems, in the example of terrorist attack to the London tube network in June 2005. However, it cannot be feasible in large-scale disasters, where also new building codes for new construction of the built environment has to be issued by the authorities before repairing.

When a disaster hits, the largest damage occurs due to mismatching or misaligned urban plans, components, or systems, or due to connections that had never been established and had been ignored. The main mistake done frequently is dealing with complex systems as if they were just complicated and separated from their environment. But elements of a system are not separated from each other, cities are not separated from their environment, and problems in one system can easily hamper to others. Therefore, solutions provided to solve a problem should be dealing with complex systems as a whole. For example, if a development plan suggests densification in an area while the disaster risk management plan would suggest the opposite, both plans need to be aligned and either the development be moved away or the disaster management be improved.

6.2 Discussion

The book has sought to know the interaction pattern among structural, organizational, tactical, and public layers, and how the outcome coming from this interaction affects the entire urban system. Organizational part consists of the upper level decision makers, tactical part consists of the personnel in the field, which could be the staff of transportation system and staff of municipalities—and public level includes lay people.

The research started with the problem definition that indicated the gap between the plans/blueprints as planned and the actual situation in the field when an incident occurs. This gap derives from the failure to consider interdependencies among system's components, secondary effects in hazard maps and social structure related to problems, such as too much confidence in current operational and technological tools, lack of experience and misunderstanding of disaster situations. The book has sought to answer the following related questions:

(1) What are the awareness and preparedness of risk levels at the organizational, tactical and public levels?
(2) What are the effects of the outcomes of decisions coming from organizational, tactical and public levels on the transportation system?

The following section will synthesize the empirical findings to answer the questions.

(1) What are the awareness and preparedness of risk levels at the organizational, tactical and public levels?

The interviews conducted during the fieldwork in London gave a broad coverage of many aspects of emergency management system in these three levels, as well as awareness and preparedness of risk at the organizational, tactical and public levels. During an emergency, policies and/or strategies are taken at the organizational level. Then staff at the tactical level follow strategies at the incident scene and have direct contact with people at the public level. The liaison occurs both vertically among these layers and horizontally within the same level. The duty of people at the organizational level is to provide adequate knowledge either to personnel at the tactical level and to people at the public level. A senior level officer from the tactical level contacts to people at the organizational levels to inform them about the situation.

The outcomes of the interviews with people from the organizational level are as in the following:

- The confidence in the current disaster risk management system is high and most of the respondents believe that the system is faultless.
- Decision makers trust to people in the tactical level. They think that the personnel is knowledgeable and know what to do during any kind of emergencies
- Decision makers and experts in the organizational level defend the idea that public information is sufficient and accessible to public. The websites of environment agency and metropolitan office and the early warning messaging service provide necessary information to public.
- The structural defences work properly and they are the last technology and sufficient to protect city from flooding.

Regarding to the first point, most of the interviewed administrators and decision makers think that the system functions always as planned without any fault. However, having faults in every system is inevitable and normal because of uncertainty of situations and scarce resources.

Furthermore, following the interviews with the emergency personnel of boroughs and Transport for London proved the second point. The existence of knowledgeable staff at the tactical level is one of the most valuable resources in London.

According to the interviews, public can easily access to information that is available in the resilience reports published online, websites and flood information messaging services. However, during the research period in London, the author has not encountered with a study on public's awareness of information in London. Moreover, having information available does not really mean that people know and understand this information. The example to this issue is given in the following paragraph.

As an example of dependability of structural defences, the flood events in 2012 in the UK can be considered. According to data released by the MET office (Meteorological Office), the total rainfall for the UK during 2012 was 1,330.7 mm, which is the second wettest on record in the UK following the year 2000.[1] Consequently, in 2012 the Environment Agency reported 200,000 warnings to households and businesses. However, in some parts of the UK, the real situation was not as planned. Embankments collapsed in Great Western Canal, and in Kempsey 1 year old £1.7 million flood defence system failed during torrential rain due to an electrical fault.[2] As a result, houses flooded and transportation in the area suspended. Inauguration of the defences in Kempsey celebrated in July 2011 by the involvement of local people. The local people had been told the state of the art on flood defences would protect them, and therefore, due to the false sense of security provided by these defences, the community had not taken any action to protect itself and its property.[3] Another major challenge dealing with low probability events is the underestimation of the size of the event that may occur. As Alan Crockford a landlord in the village reports it in the Dailymail published in 26 November 2012:

> We were given warnings from Floodline, and they just said, like, to take care; we were never given any firm warning that you had to evacuate and no-one knew to expect it this bad.

The outcomes of the interviews with people from the **tactical level** are as in the following:

- Staff of the transportation system is knowledgeable. Personnel know what to do and how to do. They know their resources and how to use them.
- Their main problems are related with the structural deficiencies in the tube stations and lines. The system is old, platforms are narrow, and some of the tube stations are very deep, and if the stairs are needed during emergencies, it is quite challenging for some of the customers to use the stairs.
- The personnel in the boroughs are also very knowledgeable. But their problems are limited resources and not interacting with other departments within the borough. For instance, when preparing emergency plan they do not work with spatial and transportation planning teams.
- There is no interaction among boroughs. They do not know each other's emergency management plan. The legal border between boroughs is another problem. A river as a border between two boroughs and flooding in this river and having two different emergency procedures could create problems during immediate response phase to flooding.

---

[1] Source: http://www.metoffice.gov.uk/news/releases/archive/2013/2012-weather-statistics, browsed in 4 January 2013.

[2] Source: http://www.dailymail.co.uk/news/article-2238532/UK-floods-wreak-havoc-houses-thought-protected-state-art-1-7million-flood-defences.html, accessed on 26 November 2012.

[3] Source: http://www.dailymail.co.uk/news/article-2238532/UK-floods-wreak-havoc-houses-thought-protected-state-art-1-7million-flood-defences.html, accessed on 26 November 2012.

Results of the questionnaires that were conducted with the staff of Transport for London and personnel of the emergency management department of boroughs confirmed the result coming from interviews with the people at the organizational level. The staff is capable to deal with emergencies and very well trained. The same can be said for the emergency personnel of boroughs. One of the main problems observed at the tactical level is not having horizontal relations with other departments. For example, while emergency department indicates an area as a first level flood risk zone, development department of the same borough opens that area to development with some precautions. Furthermore, there are couple of flood hazard modelling tools, and developers choose the one that best serves to their interest. According to the head of the emergency staff of one of the boroughs, there should be only one flood hazard model to be used legally and mandatory also for developers. Moreover, preparing an emergency plan in each borough is an obligation. At this point two problems have been identified. The first one is each borough is free to organize their budget according to their priorities. As a result, emergency plans differ significantly in quality. Second, there is no interaction among boroughs while preparing the emergency plans.

The outcomes of the interviews with people at the **public level** are as in the following:

- There is a large gap between actual awareness of information at the public level and what decision makers think about public's awareness of risk information.
- If the respondent's both economic and educational levels are higher the willingness of learning more information about how to increase their preparedness level against flood risk is higher as well.
- If the respondent's both economic and education levels are lower, though at the beginning they accept to answer the questionnaire they lose their attention in the further steps where they start not understanding the questions.
- Trust in official information providers is high at the public level.

When the results of the first and the third levels are compared, the large gap between awareness of information of public and what decision makers think about public's awareness of information became visible. The results of the survey conducted with the local people in Lewisham and Canary Wharf were totally opposite than sayings of people at the organizational level. Moreover, the results showed that majority of the respondents do not know the Environment Agency, the meaning of the flood signs and the function of the Thames Barrier. However, trust in information providers is very high.

(2) What are the effects of the outcomes of decisions coming from organizational, tactical and public levels on structural system?

This three-levelled survey approach was quite supportive to prove the gaps among levels that lead to produce different outcomes than intended. Besides, the interviews have provided insights to understand possible indirect effects of a hazard on the structural system.

6.3 Conclusion

In their studies Hale and Heijer (2006, p. 139) state that fragmented problem solving impedes to see the entire system and to get an integrated view of all types of safety requirements which regard passengers, workers, tracks, level crossings etc. In addition, Hollnagel (2006, p. 12) states the difference between "normative and normal" structure of the system. A normative system consists of laws, regulations and policies. In London, the normative system was established carefully by considering previous experiences and current hazard analysis. However, exaggerated confidence in the system hinders the differentiated normative and normal systems. At this point, maps and conceptual frameworks, which consider induced hazards and human failures as well as the hazard itself, help to reflect normal situations in addition to regulated ones. Resilience of a system can be improved by enhancing flexibility of systems and ability to cope with unexpected situations.

The most important lesson from the research is that, a disaster risk/emergency management system is constructed by rules and regulations, which is technically correct, however, failures or incidents during an emergency are emergent phenomena (Hollnagel 2006, p. 12). That kind of failures and reactions of a system, which includes both physical and social components, are hard to predict. Besides, outcome of actions, which are defined in the plan by regulations, could be different than anticipated due to constantly changing environment during disasters. In such a situation new decisions, which are not defined in the plan, have to be taken with limited knowledge of the current situation. Almost four decades ago Ian Mitroff joined the term of "the error of the third kind" or "solving the wrong problem" (Mitroff 1974, cited in Hollnagel et al. 2006, p. 22). This occurs when people insist on applying the written plans on fluctuating circumstances instead of re-organizing strategies and priorities. During crisis situation flexible systems could be better than too much order in terms of enhancing resilience.

Quick response, ability to monitor and to adapt the plan according to changes and to reorganize resources in an environment changing instantly are the keys for enhancing resilience.

References

Hale A, Heijer T (2006) Is resilience really necessary? the case of railways. In: Hollnagel E, Woods DD, Leveson N (eds) Resilience engineering: concepts and precepts. Ashgate Publishing Ltd, Hampshire

Hollnagel E (2006) Resilience the challenge of the unstable. In: Hollnagel E, Woods DD, Leveson N. (eds) Resilience engineering: concepts and precepts. Ashgate Publishing Ltd, Hampshire

Hollnagel E, Woods DD, Leveson N (eds) (2006) Resilience engineering: concepts and precepts. Ashgate Publishing Ltd, Hampshire

Mitroff I (1974) On systemic problem solving and the error of the third kind. Behav Sci 19(6):383–393

Appendix A
Questionnaire to Employees of TFL

Questionnaire to employee of TfL
Questionnaire no:

Interview Declaration
I confirm that I have carried out this interview with the named person in the named metro station and that I asked all the relevant questions fully and recorded the answers in conformance with the survey specification and within the MRS Code of Conduct and the Data Protection Act 1998. I confirm that all the information I collect will be kept in the strictest confidence, and used for research purpose only. It will not be possible to identify any particular individual or address in the results.

Funda Atun

Location .... MALE / FEMALE

Date........................Day...................... Length......................

Working status of respondent

☐ Part-time (9-29 hours)　　☐ Full time (30+ hours)

Age of respondent　　☐ 16-24 ☐ 25-34 ☐ 35-44 ☐ 45-60 ☐ 61+

How Long have you been worked at this tube station?

☐ Less than a year　　　　　　☐ 10 years and up to 20 years
☐ 1 year and up to 3 years　　☐ 20 years and more
☐ 3 years and up to 5 years　　☐ Do not know
☐ 5 years and up to 10 years

Q1. How often are you dealing with emergencies?

☐ Daily
☐ Once a week
☐ Twice weekly

F. Atun, *Improving Societal Resilience to Disasters*, PoliMI SpringerBriefs,
DOI: 10.1007/978-3-319-04654-9, © The Author(s) 2014

☐   Once a month
☐   Twice a month
☐   Couple of times a year
☐   I have never experienced

Q2. London is prone to risk of …

☐   Flooding (water overflowing from rivers, canals or streams, or rainwater/melting snow running off gardens and pavements, and overflowing from drains into the station – not burst of pipes)
☐   Terrorism
☐   Snow Storm – Bad weather condition
☐   Drought
☐   Epidemics
☐   Fire
☐   Climate Change
☐   Technological risk

Q3. How likely or unlikely do you think it is that the underground station will be flooded?

☐ Next 12 months ☐ Next 5 years ☐ Next 30 years ☐ Next 50 years

Certain to be flooded	X X X X
Very likely	X X X X
Fairly-unlikely	X X X X
Very unlikely	X X X X
Certain not to be flooded	X X X X
Don't know	X X X X

Q4. Have you ever been in a drill exercise?

☐ YES   ☐ NO

Q5. How often do you exercise a drill?

☐   Once a week
☐   Twice weekly
☐   Once a month
☐   Twice a month
☐   Couple of times a year
☐   I have never experienced

Q6. What is your satisfaction level from these exercises?

☐   Very Satisfied

☐   Satisfied
☐   Neither satisfied, nor dissatisfied
☐   Dissatisfied
☐   Very dissatisfied
☐   Don't know

Q7. How was your experience?

Q8. What is the procedure of dealing with emergencies?

Q9. What are the main problems during an emergency?

Q10. What is the procedure in case of flooding?

Q11. How does public respond to an emergency?

Appendix B
Questionnaire to the Public in London

Questionnaire to public
Questionnaire no:

Interview Declaration

I confirm that I have carried out this interview with the named person in the named metro station and that I asked all the relevant questions fully and recorded the answers in conformance with the survey specification and within the MRS Code of Conduct and the Data Protection Act 1998.

I confirm that all the information I collect will be kept in the strictest confidence, and used for research purpose only. It will not be possible to identify any particular individual or address in the results.

Funda Atun

Location .... MALE / FEMALE

Date........................Day...................... Length.....................

Working status of respondent

☐ Part-time (9-29 hours)   ☐ Not working/looking after house
☐ Full time (30+ hours)   ☐ Not working – invalid/disabled
☐ Not working – unemployed   ☐ Student
☐ Not working – retired   ☐ Other..................................

Age of respondent   ☐ 13-18 ☐ 19-24 ☐ 35-44 ☐ 45-60 ☐ 61-75 ☐ 75+

Respondent is   ☐ Chief income earner ☐ not chief income earner

What is the number of households in your home? (Including yourself)

Children aged 9 and under	☐ YES	☐ NO
Children aged 10 to 17	☐ YES	☐ NO
Adults aged 18 to 69	☐ YES	☐ NO
Adults aged more than 70	☐ YES	☐ NO

F. Atun, *Improving Societal Resilience to Disasters*, PoliMI SpringerBriefs, DOI: 10.1007/978-3-319-04654-9, © The Author(s) 2014

Which of these best describe your home?
☐ Basement flat
☐ Caravan
☐ Detached house
☐ Ground floor flat
☐ Ground floor maisonette
☐ Semi-detached house
☐ Terraced house
☐ Other

Is the property in sight of a potential flood source?
☐ YES ☐ NO ☐ DO NOT KNOW

How long have you been in this address?
☐ Less than a year
☐ 1 year and up to 3 years
☐ 3 years and up to 5 years
☐ 5 years and up to 10 years
☐ 10 years and up to 20 years
☐ 20 years and more
☐ Do not know

Which, if any, of these applies to you?

☐ I own a mobile phone
☐ This household receives digital TV/Radio
☐ This household uses broadband internet
☐ Members of these household uses internet outside home for personal use
☐ I read local newspapers on a regular basis

PART 1: PERCEPTION AND AWARENESS OF RISK CONDITION
Q1. Please name the likeliness of these disaster types according to your judgment between 1 to 10 where 1 is the most probable and 10 is the least.

☐ Flooding (all kind of flooding)   ☐ Climate change
☐ Terrorism   ☐ Technological risk
☐ Snowstorm   ☐ Earthquake
☐ Drought   ☐ Ash crisis due to volcanic
☐ Epidemics   eruption in another place, such as
☐ Fire   Iceland

Q2. Is your HOME or WORK or SCHOOL at risk of flooding?

Home	A great deal	A fair amount	Not very much	Not at all
Work	A great deal	A fair amount	Not very much	Not at all
School	A great deal	A fair amount	Not very much	Not at all

Q3. How likely or unlikely do you think it is that your HOME (H)or WORK (W) or SCHOOL(S) will be flooded?

	☐ Next 12 months	☐ Next 5 years	☐ Next 20 years	☐ Next 50 years
Certain to be flooded	H W S	H W S	H W S	H W S
Very likely	H W S	H W S	H W S	H W S
Fairly-unlikely	H W S	H W S	H W S	H W S
Very unlikely	H W S	H W S	H W S	H W S
Certain not to be flooded	H W S	H W S	H W S	H W S
Do not know	H W S	H W S	H W S	H W S

Q4. Are you worried about the possibility of your HOME/WORK/SCHOOL being flooded?

☐ Very worried ☐ Worried ☐ Not very much ☐ not at all

Q5. How did you first find out about the risk of flooding of your HOME/WORK/SCHOOL?

....

PART 2: AWARENESS OF FLOOD WARNING

Q6. Do you know the meaning of these signs?

Q7. Have you ever received a flood warning? ☐ YES ☐ NO

Q8. If the answer is yes, how did you receive the flood warning, and when?

....

Q9. Do you rely on the authorities when the river Thames is flooding?

....

PART 3: AWARENESS OF THE ACCESS OF INFORMATION

Q10. Have you registered with flood warning messaging service?
☐ YES   ☐ NO

Q11. Do you know the environment agency's flood warning website?
☐ YES ☐ NO

Q12. If the answer is yes, how often do you check the website?
☐ Often ☐ Sometimes ☐ Rarely ☐ Bad weather conditions ☐ Never

PART 4: AWARENESS OF INFORMATION PROGRAMMES

Q13. Are you aware of any information programmes on flood response / improving conditions of home in case of flooding?
....

PART 5: POPULATION'S INDIVIDUAL PREPAREDNESS

Q14. Have you undertaken any flood prevention measures at your HOME/WORK/SCHOOL?
....

Q15: Do you have any of the followings to protect your home from flooding?

YES	NO	Insurance
YES	NO	Sandbags
YES	NO	Raised the property off the ground
YES	NO	Flood boards / flood gates / airbrick covers
YES	NO	Water pumps
YES	NO	Walls around the property
YES	NO	Flood proof exterior walls

Q16: Have you undertaken any of the following actions?

Checking the level of river	Very often – often – sometimes – never – depending on the weather
Checking the flood warnings	Very often – often – sometimes – never – depending on the weather
Checking the weather forecast	Very often – often – sometimes – never – depending on the weather

Q17. Do you know what to do during an emergency? Do you have a plan? If yes please explain.

☐ YES ☐ NO...

Q18. Have you ever informed by the authorities about the evacuation plan?

☐ YES ☐ NO...

Q19. Do you know where are you going to evacuate? ☐ YES ☐ NO

(If yes)Where?...

Q20. If you need to evacuate, are you going to use which vehicle?

☐ Your own car ☐ Neighbour's/friend's car ☐ Public bus–train–boat

☐ Vehicle provided by the government

Appendix C
The Results of the Questionnaire

Respondent Information

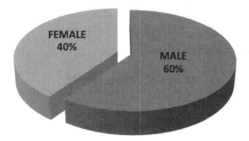

Graph C.1 Gender, Lewisham

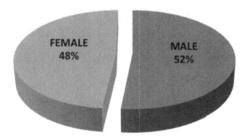

Graph C.2 Gender, Canary Wharf

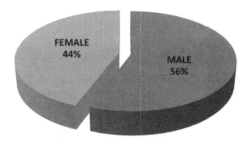

Graph C.3 Gender, Lewisham + Canary Wharf

F. Atun, *Improving Societal Resilience to Disasters*, PoliMI SpringerBriefs,
DOI: 10.1007/978-3-319-04654-9, © The Author(s) 2014

Working status of respondents

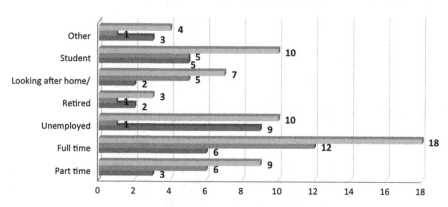

Graph C.4 Working status of respondents

Age of respondents

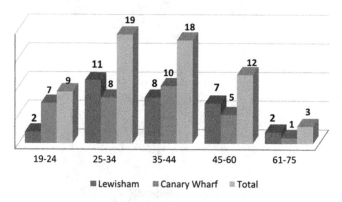

Graph C.5 Age of respondents

Graph C.6 Chief income
earner or not

Chief income earner or not

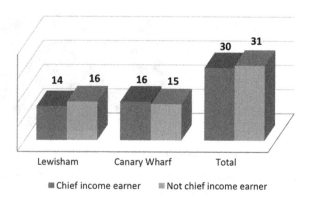

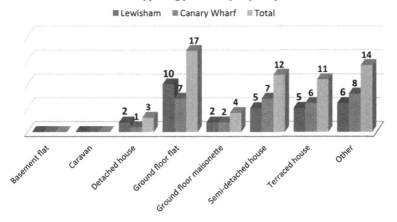

Graph C.7 Typology of the property

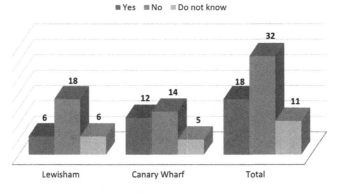

Graph C.8 Whether the property in sight of a potential flood source

How long have you been in this address?

■ Lewisham   ■ Canary Wharf   ■ Total

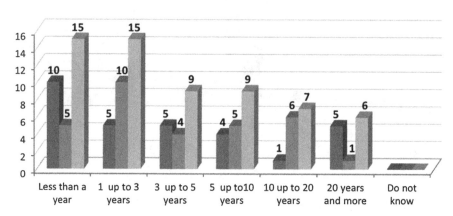

Graph C.9 Duration of residing at the specified address

Which, if any, of these applies to you?

■ Lewisham   ■ Canary Wharf   ■ Total

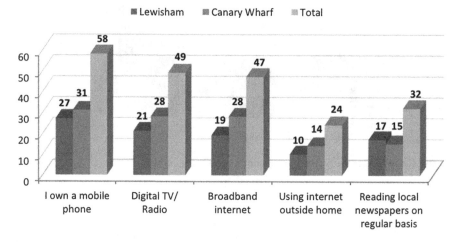

Graph C.10 To use technology

Household numbers of the respondents

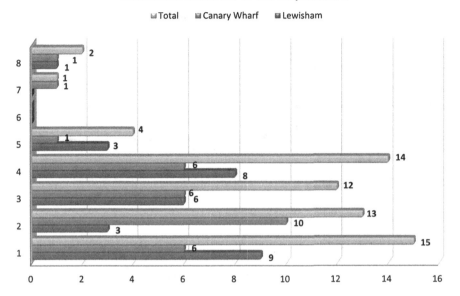

Graph C.11 Household numbers of the respondents

Table C.1 Perception and awareness of risk conditions Please name the likeliness of these disaster types according to your judgment between 1 and 10 where 1 is the most probable and 10 is the least

	Lewisham	Canary Wharf	Total
Snow storm	2	1	1
Terrorism	1	3	2
Fire	4	2	3
Climate change	3	7	4
Technological risk	6	5	4
Flooding	5	6	6
Epidemics	7	4	7
Drought	7	8	8
Ash crisis	9	9	9
Earthquake	10	10	10

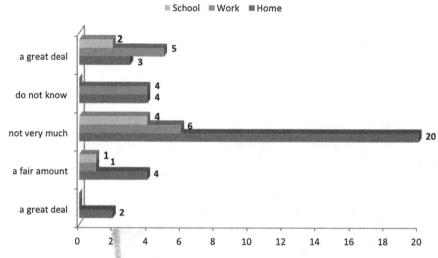

Graph C.12 Home/work/school at risk of flooding—Lewisham

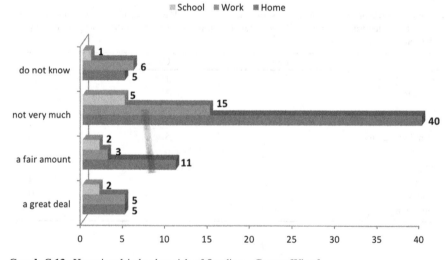

Graph C.13 Home/work/school at risk of flooding—Canary Wharf